New Industry Creation:

Discovery to Innovation

Murali Nair

"Great is the human who has not lost his childlike heart."
- Mencius (Meng-Tse), 4th Century BCE

Disclaimer:

The views expressed in the book are those of the author and do not necessarily represent the views of the National Science Foundation or the United States.

For my beloved mahalko

Contents

Preface   xi

Introduction   **1**

> *State of Technology*
>
> *Technology Innovation and Small Firms*
>
> *Manufacturing, a Strategic Sector*

Part I

General Considerations

Chapter 1   Society and Technology   **17**

> *Nature of Technology*
>
> *Role of Technology*
>
> *Technology Evolution*
>
> *Technology and Its Discontents*

Chapter 2   Innovation Economy   **35**

> *Innovation Economics*
>
> *Technology Innovation*
>
> *New Job Creation*

Part II

Discovery to Innovation

Chapter 3   Discovery Science   **55**

> *Federal Government as Partner*
>
> *Federal Basic Research Funding*

Chapter 4   **Innovation Research**   **75**

Federal Innovation Research Funding

Chapter 5   **Infrastructure**   **91**

Pillars and Partners

Chapter 6   **Engender Demand**   **124**

Overarching Considerations

Facilitate Market Entry

Go-to-Market Strategies

Chapter 7   **Consequence**   **148**

Canonic Market Pathways

Probable Outcomes

Possible Impact

Part III

New Industry Creation

Chapter 8   **Make in America**   **164**

State of American Manufacturing

Federal Role

Economic and Societal Implications

Future of Manufacturing in America

Chapter 9   New Industry Creation   **186**

Evolution of Discovery-based Innovation

Barrier to Innovation

Future of Work, Future of Startups

New Industry Creation

Acknowledgement   216

List of Tables   217

Photo/Illustration Credits   218

Startup Profiles Listing   219

References   224

Index   231

Preface

Mere power and mere knowledge exalt human nature but do not bless it
- Francis Bacon

This is the first volume in a trilogy on technology and humankind that seeks a deeper understanding of the role and impact of technology, the conditions and standards of life it helps create, the contents and discontents of technology in human society and in our world, and towards an enhanced level of human consciousness, to spiritual machines and godlike human beings. Our world is rapidly changing. Humanity grapples with multiple challenges on a grand scale including those related to planet Earth's existential questions, and concerns around demographics, people's aspirations, poverty, and human and environmental degradation. The drivers in this century are globalization, population increase, poverty and the politics of resentment and depravation, the rise of new powers, the decay of international institutions, climate change, and the geopolitics of energy.

Simultaneous revolutions – knowledge, political, and cultural - along with the intellectual, social and economic edifice, lead some to consider the present the best era in human history, with progress along many and multiple fronts. One impact will be to further accelerate the decoupling of gains in productivity from gains in the standard of living for the middle class. Exchange productive activity for income to meet essential needs – this basic premise at the heart of human societies is beginning to be radically transmuted. In the past, improvement in economic efficiency has generally led to improvements in wages for the majority. Now the fundamental relationship between technology and employment is being transformed, and nearing a threshold beyond which so many jobs are lost that the level of consumer demand falls below the level necessary to sustain healthy economic growth. We are poised for another wrenching transition with the current ongoing loss of manufacturing jobs. Should anyone work at all; perhaps few humans need to work. Machines and robots can perform essential work altering the very meaning and essence of work, and may result in the end of work.

Prevailing economic theory suggests that a country's job structure and income distribution are determined more by its own domestic policies - education, investment, taxes - plus shifts in technology than by anything its competitors do. Productivity figures typically rise as a result of outsourcing and robot deployment. With the hemorrhaging of jobs to intelligent machines, the fundamental role of labor in the economy of the future is being questioned, creating a shift in the relative value of technology and capital to labor. The last ten years the only decade since the Great Depression when there have been no net jobs added to the U.S. economy while along with productivity gains corporate profits soared and unemployment barely declined. Major dislocations in the role of capital as a reliable and efficient factor of production, has led to a reduction in income for middle-income wage-earners having a noticeable impact on aggregate demand particularly in consumer-oriented societies. While both labor and capital have been globalized, the bulk of consumption in the world economy remains in wealthy industrial countries resulting in a mismatch between the distribution of income and the central role of consumption in driving global economic growth.

Economic growth, faith in business, and our confidence in technology must all urgently come to bear to create millions of sustaining, high-wage jobs each year. Throughout U.S. history, government has played a far more active role in fostering the creation of new firms and new industries than market-oriented thinkers like to admit. Most of modern technology - jet planes, satellite communications, integrated circuits, computers, and the internet - that has been driving the U.S. economy did not appear spontaneously from market forces. There has been too little recognition that the U.S. government has developed a remarkable capacity to nurture innovation and the commercialization of new technologies across all economic sectors. Through financing, guidance, and connecting key players, government agencies have played a crucial role in developing America's innovation economy.

There exist innumerable instances of U.S. public-sector seeded industries although now many products invented in America are mostly made abroad. Both laissez-faire and strategic industry policies are required to seed sunrise industries. In critical technology areas, the venture capital model is broken. Banks, hedge funds, and venture capital firms are geared toward investing in financial instruments and software companies. In such endeavors, even modest investments can yield quick and large returns. Financing brick-and-mortar factories is expensive and painstaking, and offers far less potential for speedy returns. The need for creative new financial instruments for technology-specific funding mechanisms is acute.

We should all be enamored by innovation. It is the modern affliction. From now on innovation thinking is required of all - it is the new economic order. Research and innovation is the spark for creating whole new industries, to manufacture in America, and to form the catalytic high-technology job creation engine. How to create tens of millions of well-paying, fulfilling, ecofriendly, and sustainable jobs in America? Institutions that are critical to innovation must be supported by expanding funding not just for university research, but for the kinds of mechanisms and institutions that help foster commercialization of research through viable business models. Many important innovations present serious challenges in mass production or mass distribution. Others require considerable infrastructural investment or coordinated investments by many different entities. The danger is that other nations through technology diffusion, spillovers, and usurpation of intellectual property reap the benefits of U.S. investments in science and technology. More effective government policies incorporating externalities, scale economies, retainable industries, and filling-in of hollowed-out supply-chains, can keep the industries of the future from moving offshore. Use of the federal government's coordinating abilities to overcome information mismatch problems to bring together innovators and markets is critical.

This volume details a possible framework to successfully transform public investments in basic and applied research to economic value and societal impact through federal funding of technology startups that facilitate innovative business models through coordination and partnership with the private sector. It aspires to effectively deploy the triple helix of private equity, academia, and industry to ignite local economic development notwithstanding the fact that it is the job of business to create jobs not the government that can only help seed new industries.

For over three decades, the small business innovation research program administered across several federal agencies has provided funds to thousands of technology startups, and helped them bring new products and services to market. This program nurtures innovation for economic and societal benefit by investing in industry-transforming science and engineering research and development. By promoting collaboration between industry and academe it has helped create successful technology startups and increased long-term, leading-edge U.S. innovation capacity. The mission is to assist such federally-funded startups in their technology development efforts, and where applicable, to manufacture in America - born here, make here, and sell everywhere. To seed high-wage jobs and future industries across the entire technology spectrum, public-private partnerships can combine federally funded early-stage technologies, private venture capital funds, a web-based platform to facilitate strategic alliances, powerful technology-specific nationwide networks of partners and investors, and opportunities for startups to leverage under-utilized U.S. manufacturing infrastructure.

Over four dozen profiles of high-tech startups, all recipients of grants from this taxpayer-funded program covering the entire gamut of technologies arising out of basic research investments, from life sciences to physical sciences, from synthetic biology and medical devices to nanotechnology and sophisticated instruments, are sprinkled throughout the text. These feature the different and varying aspects of current technology development trends such as for instance the interplay of materials, biology, and information technology innovation. In the future, one laudable goal would be to match this program's estimated average annual public investment of approximately $2.8 billion with an equal amount in private investment in these federally-funded technology startups. A complementary stretch goal can be to create, over the long-term, a private hundred billion dollar endowment using foreign direct investment, and private and public sources, to invest in next-generation industries that will enable the creation of two million high-skill jobs every year, especially targeted to capital intensive products and strategic manufacturing in America. This will entail new financial models such as bonds, prizes, derivatives, and maybe even the state staking equity positions in funded startups. It may also require new policies and the modification of current innovation practices to facilitate additional private funding, in order that we may derive more powerful economic outcomes.

Murali Nair
December 31, 2015
Stafford, Virginia
muralinair@yahoo.com

Introduction

Will humanity find new ways of using its labor, or will the future be given over to forced leisure, is a matter of much debate among economists. Enormous challenges and opportunities lie ahead. It is opportune to meditate on the relation between individual liberty and government authority, and the ever-expanding possibility generated by the division of labor. In economics, the price system is a mechanism for coordinating knowledge. Competition is key, where chance and good luck are often as important as skill and foresight in determining the fate of different people[1]. Globalization – the removal of barriers to free trade and the closer integration of national economies – can be a force for good and has the potential to enrich everyone in the world. It is spreading not just trade and technology but also resentment and impatience[2]. Millions of great minds that might otherwise have toiled at subsistence farming can instead now join the modern economy and share the burden of knowledge with rich-world researchers – a sharing that information technology makes ever easier. Today's environmental reality is linked powerfully with other realties, including growing social inequality and neglect, and the erosion of democratic governance and popular control. How can the operating instructions for the modern world economy be changed so that economic activity both protects and restores the natural world?

---

[1]Hayek, Freidrich – The Road to Serfdom, 1944

[2]Stiglitz, Joseph - Globalization and Its Discontents, 2003

Capitalism as constituted today, with an unquestioning society-wide commitment to economic growth at almost any cost, produces an economic and political reality that is highly destructive of the environment. Our gas-guzzling, water-wasting, plastic-producing civilization is not sustainable. America's power includes both military might and the "soft power" of language, culture, commerce, technology, and ideas. There are interesting parallels with ancient power centers: a focus on the capital city, reliance on military instruments, privatization of public goods, parochial attitudes towards the outside world, problems with borders, and growing complexity. We can learn lessons for America today: instill greater appreciation of the wider world, stop treating government as a necessary evil, fortify the institutions that promote assimilation, take some weight off the military, and focus on priorities - climate change, infrastructure, education, trade barriers, malnutrition, conflicts, financial instability, human migration, sanitation and communicable diseases, poverty, and corruption.

Institutions have become too rigid to accommodate truly revolutionary changes. The quest for a domestic policy on the scale of the planet is probably relevant[3]. We have enormous investments in technologies designed with little regard for the environment, together with powerful corporate interests, whose overriding objective is to grow by generating profit, including profit from avoiding the environmental costs they create. We have markets that systematically fail to recognize environmental costs unless corrected by the government, a government that is subservient to corporate interests and the growth imperative. There is rampant consumerism spurred by a worshipping of novelty and by sophisticated advertising, economic activity so large in scale that its impacts alter the fundamental biophysical operations of the planet. These combine to deliver an ever-growing world economy that is undermining the planet's ability to sustain life. How do we harness economic forces for sustainability and sufficiency? The creativity, innovation, and entrepreneurship of businesses operating in a vibrant private sector are essential to designing and building the future[4].

State of Technology

Scientific revolutions are the tradition-shattering complements to the tradition-bound activity of normal science. Immense effort and ingenuity have been required to bring nature and theory into closer and closer agreement. In the sciences fact and theory, discovery and invention, are not categorically and permanently distinct. First we have

[3]Habermas, Jurgen – The Philosophical Discourse of Modernity, 1985

[4]Speth, James – The Bridge at the Edge of the World: Capitalism, the Environment, and Crossing from Crisis to Sustainability, 2008

discoveries or novelties of facts, and then inventions, or novelties of theory[5]. Research is a strenuous and devoted attempt to force nature into the conceptual boxes supplied by professional education. An excessive concern with useful problems, regardless of their relation to existing knowledge and technique, can so easily inhibit scientific development. Such marvelously adapted organs as the eye and hand of a human were products of a process that moved steadily from primitive beginnings but toward no goal. Unlike the engineer, and many doctors, and most theologians, the scientist need not choose problems because they urgently need solution, and without regard for the tools available to solve them. Science has reached its present state by a series of individual discoveries and inventions that, when gathered together, constitute the present body of technical knowledge.

The modern age has belonged to material progress and its predominant source has been science. Technology is deeply embedded in our natural history. Technologies are always both constructive and constructed by historical, social, and cultural contexts. Revolutions in science have often been preceded by revolutions in measurement. Technology has often played a vital role in the emergence of new sciences especially in fields like medicine and instruments, when the principal reason is an external social need. Part of our difficulty in perceiving the profound differences between science and technology relate to the fact that progress is an attribute of both fields. Important new technologies are described as "extensions" of basic human capacities[6] but new and powerful forms of artificial intelligence represent an extension of the dominant and uniquely human capacity to think, fundamentally different from any other technological extension. Transformational waves of new technologies ripples through the economy and creates new opportunities and wealth. Everything we know about future industries and technologies suggests that they will offer ever-greater rewards to flexibility, openness, reinvention, crowd-sourcing, and all other manifestations of individuals and groups keenly attuned to their surroundings. American society should be hospitable toward those traits, and should foster them better and more richly than other societies can[7].

Everything from cars to carpets could ultimately have social connectivity embedded in it with the internet of things, and the intercloud – a cloud of clouds in the way that the internet is a network of networks. The powerful triplet of information and communications

---

[5]Kuhn, Thomas - The Structure of Scientific Revolutions, 3rd Edition, 1996

[6]McLuhan, Marshall – The Mechanical Bride: Folklore of Industrial Man, 1951

[7]Note: In each chapter, snapshots of publicly-funded startups are featured to showcase the variety of technologies supported. The underlying theme for the five startups presented in this chapter is how such fledgling companies can and do provide solutions to global challenges such as health, education, energy, water, and lighting respectively.

technologies; biotechnologies and neuroscience; and renewable energy technologies, is just at its beginnings. Products are becoming like services, a product is merely a service waiting to happen[8]. It is only a matter of a decade or so before exponentially evolving

---

No Need to Wait for an Organ Transplant!
Cardiac-derived Patch for Heart Disease

This startup based in Eden Prairie, Minnesota is being provided federal funding in the amount of approximately $1.0 million for research and development work from July 2013 to October 2017 to support the continued development of using proprietary perfusion decellularization technology to create a fully revascularized cardiac patch for the treatment of ischemic heart disease and congenital heart repair (see notes below). Current surgical approaches for cardiac reconstruction utilize synthetic materials that do not have the ability to grow and remodel with the patient. Feasibility is demonstrated by in-vitro (in the lab) and in-vivo (within a living organism or natural setting) characterization to create a cardiac patch to promote faster reconstruction of functional tissue by providing a fully perfusable scaffold with a composition and architecture similar to native cardiac tissue. This product will have significant advantages over existing technologies, including full thickness, biological, cardiac-derived matrix material; vascular supply to support migrating cells and remodeling; superior mechanical properties; and obviating immunosuppressive therapies. This life sciences company is developing technology to remove cells from animal/human organs and tissues thus facilitating the creation of numerous products, including the potential of fully biological transplantable human organs. Their decellularization process results in a scaffold or biological matrix that retains the shape and appearance of the source organ or tissue, but with an opaque appearance. The original architecture, mechanical properties and vascular network of the anatomical structure are also maintained, which is critical for the cellular interactions that facilitate the recellularization of the scaffold and enable the growth of new tissue or a new functional organ. In addition to the company's internal efforts focused on the development of its first product – a biological mesh for hernia repair – scientists around the globe are working with its technology and are taking critical steps toward the creation of human organs including the liver, lung, kidney, pancreas and heart. While their biological mesh development program has been conducted internally, the intent is for current and future products to be developed both by the startup and in partnership with external groups already working with its technology in order to utilize this expertise to develop individual products in the most efficacious, cost-effective and timely way possible.

Notes:
Perfusion:	passage of fluid through blood vessels to an organ or tissue
Decellularization:	remove cells from
Revascularized:	surgically improve blood circulation
Ischemic:	local deficiency of blood supply

---

information and communications technology will allow human interaction with essentially any degree of fidelity we wish, as digital technology continues to increase in power a thousand-fold each

---

[8]Rischard, Jean-Francois – High Noon: Twenty Global Problems, Twenty Years to Solve Them, 2002

decade, at some point large computer networks might "awaken" with superhuman intelligence. Or biological science may provide the means to improve natural human intellect. Sustainable, equitable development, pressing global problems, robust and productive rural economies, social entrepreneurship, micro-lending, alternate energy, solid state lighting, biotechnology, healthcare solutions, improved agricultural efficiency and productivity, infrastructure improvements, communications, alternate technology development, transportation, water recycling, the generation, conservation, recovery, harvesting and conversion of energy, energy innovations, all call for research breakthroughs, an entrepreneurial cadre, and technology innovation. A third revolution is underway – the convergence of the life sciences and

---

Martha Madison's Marvelous Machine
New Learning Pathways

This startup based in Pittsford, New York is being provided federal funding in the amount of approximately $1.0 million for research and development work from August 2012 to June 2017 to develop, test and commercialize a collaborative educational game aimed at a broad spectrum of middle-school physical science students that will improve educational outcomes and increase interest in science, technology, engineering and mathematics. Martha Madison's Marvelous Machine, an educational game based on pedagogical best practices, correlated to curricular standards and supported by lesson plans, challenges students to work as a team to solve progressively harder physical science problems. Sophisticated scaffolding guides learners and helps them achieve their full potential by encouraging exploration yet minimizing frustration without giving the answer away. By tracking play mechanics, the game will test for student engagement, enhancement of scientific inquiry skills and knowledge of physical science principles. The game will cover a full year of physical science curriculum.

---

the physical sciences, the transformation of biology from a discovery science into an information science as evidenced by precision medicine where pharmaco-genetics and personalized medicine meet. We now have nano-bio-info convergence, and biomimicry, the science of making technological and commercial advances by copying natural processes. Smart systems may well be humankind's best hope for dealing with its pressing environmental problems, notably global warming. Life's essence is information. Indeed, synthetic life itself is within humanity's grasp. The age of artificial life has begun.

Companies are already developing ingestible chips that can be embedded in pills, turning them into networked, digital drugs. Industrial biotechnology using agricultural feed-stocks, rather than petroleum-based ones, to develop industrial enzymes as catalysts used to make ethanol, food, animal feed, and textiles, to produce chemicals, plastics and fuels, is a future growth industry. The boom in bio-refineries, the use of bio-plastics, and the growth of the synthetic biology industry which aims to make wholesale changes to living things, are palpable.

There are enzymes being developed to make biofuels, to improve agricultural productivity, to find cleaner energy sources, and to protect the environment. There is now demand for biologically derived materials, materials that fatten when stretched, intelligent materials, matter compilers and programmable matter, structural DNA, DNA origami, programmable nano-machines, and do-it-yourself biology. Doctors are rerouting nerves to give patients a more natural control of prosthetic arms and bring paralyzed limbs back to life. Genetically modified agriculture is now commonplace and two species, the brain of one in another species, is already being done.

Automatic, highly accurate and real-time translation between the major languages of the world will greatly lower the barriers to international commerce and collaboration. Another challenge is how to do the same things as today, but with ten to hundred times less power. The convergence of wireless communications, social networking and medicine will transform healthcare, when your carpet calls your doctor. In Japan where labor to care for an aging population is scarce, innovation in robotics is proceeding unabated. The rising cost of looking after people across the rich world will only encourage further development. We will be soon swimming in sensors and drowning in data. The maker movement, mass customization, additive manufacturing, the open hardware movement with their implicit call to banish intellectual property rights, the broken venture capital model, and new financial instruments based on crowdsourcing are recent developments.

The first commercial three-dimensional bio-printer for manufacturing human tissues and organs is already in the market. Regenerative medicine - to start with, only simple tissues such as skin, muscle and short stretches of blood vessels will be made; because the machines have the ability to make branched tubes, the technology could, for example, be used to create the networks of blood vessels needed to sustain larger printed organs like kidneys, livers and hearts. Cognition drugs are the next frontier. A bio-chip which might contain a sophisticated computer and be grown in a manner similar to the way a tree evolves from a seed. It is now possible to realistically imagine new life forms, synthetic life, synthetic biology, immortality, new ways of creating life, and human cloning. The field of epigenetics, the reverse engineering of the mind, and the growing nascent market for "biosimilars" the biotech equivalents of generics, will comprise the near future. The number of patents using DNA from sea-creatures is rocketing – marine life is a hundred times more likely to contain material useful for anti-cancer drugs than is terrestrial life. Ten of the world's twenty-five top-selling drugs are already derived from natural resources.

Technology Innovation and Small Firms

In this age of knowledge, the key strategic resource necessary for prosperity has become knowledge itself – educated people and their ideas. Unlike natural resources, such as iron ore and fossil fuels, that have driven earlier economic transformations, knowledge is inexhaustible. The more it is used, the more it multiplies and expands. Intensive growth is powered by discovery of even better ways to leverage workers and resources. This kind of growth allows continuous improvement in incomes and welfare and enables an economy to grow even as its population decreases. Economists label this "technology", the all-purpose improvement factor responsible for this growth. Sometimes the government has to create the necessary infrastructure as it did when it built the interstate highway system. There is an opportunity now for the federal government to design the twenty-first century public infrastructure that will support a dynamic and innovative economy over the next fifty years.

We can build on a record of successful promotion of research breakthroughs and innovative firms to offer up creative solutions. Both top-down public policies and bottom-up marketplace innovations must play a role in propelling the world to the post-petroleum age of carbon-free emissions. A good example is solar photovoltaics: it presents significant manufacturing challenges and expanded installation of them requires coordinated investments by producers, homeowners, installers, and electrical utilities. Federal effort is required to facilitate the rapid deployment of the new technology. One task is to help firms create the skilled labor force needed to make and install the products. Another task is to create a regulatory framework, standards, and incentives that encourage consumers and private firms to make the necessary investments. Government support works best when it is limited to the risky early stage of development. For the government to do these things well (e.g. infrastructure, basic research), it is best to get out of the way of entrepreneurs, reform the public sector, and invest wisely.

There is an urgent need to invigorate the country's eroding "intellectual infrastructure". Scientific discovery, for ages sprung primarily from the minds of singular giants, is now more likely to arise from large, distributed teams. The bar code, traffic congestion charging, the cervical pap-smear, and the internet have one common attribute - all emerged from work done at America's pre-eminent research universities. These mighty institutions are creative machines unlike any other that we have known in our history[9]. They stand at the center of America's intellectual and technological global leadership but are now under threat as never before. This is the innovation pipeline,

---

[9]Cole, Jonathan - The Great American University: Its Rise to Preeminence, Its Indispensable National Role, Why It Must Be Protected, 2010

from blue-sky research to entrepreneurial development to market, that transforms basic ideas into new economic building blocks - microchips, lasers, aircraft, telecommunications, the mapped genome and more.

One modern factor to consider is the speed of knowledge creation and the "burden of knowledge: as ideas accumulate it takes ever longer for new thinkers to catch up with the frontier of their scientific or technical specialty. Roughly a century lapsed between the first commercial deployment of James Watt's steam engine and steam's peak contribution to economic growth. Some four decades separated critical innovations in electrical engineering of the 1880s and the broad influence of electrification on economic growth. Innovations of the late nineteenth century drove productivity growth until the early 1970s. Today for medical devices, it takes about seventeen years until a new technology is fully accepted as the dominant practice even for something that is new and better.

Information innovation is still in its infancy. One of the chief ways in which intelligence presses forward is through innovation, which is now recognized as one of the most important contributors to economic growth. Innovation, in turn, depends on creative people who dream up new ideas and turn them into reality. The importance of innovation and risk-taking in the economy is exemplified by many of our best companies that were started in recessions, when necessity makes innovators even more inventive and risk-takers even more daring. We still have the most creative, diverse, open, and innovative culture, in a world where the ability to imagine and generate new ideas with speed and to implement them through global collaboration is the most important competitive advantage. Innovation is unpredictable and uncertain; a fragile balance between creativity on the one hand, and discipline and efficiency on the other.

More than ever before governments, universities, and firms together spend around $1.4 trillion a year on research and development (R&D). Poorly crafted regulation may unduly raise the cost of new research discouraging further innovation. Lawmakers must ensure that regulations and other related government policies support, not retard, innovation. The West's intellectual property system is a mess because it grants too many patents of dubious merit. Backlog in the U.S. Patent and Trademark Office is more than one million with waiting times around three years. How far have we strayed from the notion of innovation that we need to acquire patents to fend off potential lawsuits?

According to National Science Foundation data, more than fifty percent of the doctoral-level scientists and engineers who work for industry are employed at firms with fewer than five hundred employees. Small businesses employ thirty percent of high-tech workers such as scientists, engineers, and information technology professionals. Startups are playing an increasingly important role in the

U.S. economy driven by two profound forces – the desire to survive and our capacity to create. Many big companies are scrapping large, centralized R&D teams in favor of smaller in-house units and alliances

---

Rethink Energy Storage
Printable Non-lithium Battery Technology

This startup based in Alameda, California is being provided federal funding in the amount of approximately $1.1 million for research and development work from March 2013 to May 2017 to expand the performance of its novel zinc battery chemistry. The technology leverages a high conductivity polymer electrolyte, and further characterizes the battery system for small portable and flexible electronics applications. The zinc battery chemistry involves ultrathin, flexible and rechargeable battery technology that utilizes an air-stable, earth-abundant, robust, and non-lithium materials set that is easily manufactured by print-based processing and is scalable to large dimensions. The goals are to increase understanding of this new battery chemistry, demonstrate and characterize its unique flexibility, scale the technology to pilot-level manufacturing, and improve its commercially relevant performance specifications. This will help establish new battery chemistry and a manufacturing paradigm which can be disruptive to markets requiring novel device functionality and form-factors. This technology also allows for significant reduction of the cost and environmental impact of batteries. This new approach to battery manufacturing presents the opportunity to repurpose the printing industry to produce next-generation batteries. Despite considerable prior work in the field of batteries, there is a large mismatch between available battery technologies and the performance, form-factor, cost, and manufacturing requirements needed to serve as a platform battery system to power flexible and wearable electronics, robotics, sensors, energy harvesters, displays, and wireless electronics. The novel battery technology being developed can alleviate these constraints and potentially revolutionize the portable electronic market to achieve new form-factors, capabilities, and spur adoption in new application areas. The startup founded in 2010 aims to reshape the battery landscape. Currently available battery technologies limit the pace of improvement in design and functionality of portable electronic devices. The company will address these shortcomings with its breakthrough zinc-based battery technology that removes longstanding limitations on the rechargeability of such batteries and enables the production of ultrathin, flexible, high energy density rechargeable batteries for significantly lower cost and without the design limitations or safety concerns of other battery technologies. The company aims to improve portable power by significantly lowering its cost and by removing form-factor limitations and safety concerns. Battery specifications are application-driven. This motivates product development especially since many possible products do not exist today. In some ways the startup has to create markets and develop a strategy for a select few customer segments. Some customers want products to put out into the market now, others are interested in wearable electronics, and others in developing new products they can potentially design with this startup's technology – these often lead to educating potential customers to various possibilities. Raising capital to finance a battery technology company is difficult and takes time – investors, more comfortable with software startups, are wary of battery companies. The well was poisoned too - battery companies had gone bankrupt, there was investor-prestige risk, to them hardware and materials startups posed too much risk.

---

with promising startups. Since the early 1980s, the federal government has funded approximately four thousand high-technology small businesses a year across the country. How many among the R&D 100 Awards over the last forty years have led to successful small firms and how many jobs and how much profits were created by these firms? The results show that these federal government-nurtured firms consistently account for a quarter of all U.S. R&D 100 Award winners - a powerful indication that these federally-funded startups have become a key force in the innovation economy of the United States.

Water is the New Oil
Deep-ultraviolet Technology for Water Disinfection

This startup based in Charlotte, North Carolina was provided federal funding in the amount of approximately $1.1 million for R&D work from January 2009 to June 2014 to bring to market a low-power, point-of-use water disinfection system designed to retrofit into existing non-germicidal filtration systems. The technology uses ultraviolet (UV) light emitting diodes (LED) along with a novel and proprietary flow-cell design. Current UV point-of-use water disinfection uses discharge lamps that are bulky and fragile with a relatively large form-factor. They require high voltages, and their disposal is environmentally problematic. The use of UV LEDs will instead allow a smaller form-factor, and be able to function at lower overall electrical power levels. LEDs are potentially 4-5 times more energy efficient than current UV bulb technology. The product developed under this R&D program is a first of its kind providing a point of entry for ultraviolet LEDs into the large point-of-use water sterilization market. The low-power aspect and small form-factor flow-cell makes the system potentially suitable also for battery-operated field applications such as military or medical field operations where line voltage is not available. This startup addresses the need for clean water with a low-cost, small-footprint point-of-use water disinfection system.

Recently, 24% of U.S. community water systems violated safe drinking water standards for microbes one or more times. The use of chlorine has been shown to generate a byproduct that has been linked to reproductive health issues. In addition to concerns about water safety, continued water shortages are providing a significant threat to global economies. The demand for water continues to escalate at unsustainable rates. Globally, water consumption is doubling every twenty years. By 2025, it is estimated that about one-third of the global population will not have access to adequate drinking water. Shortages are driving rapidly increasing prices for water. Reuse and point-of-use treatment of water offers one of the best methods for combating water shortages. It is estimated that the average household can reuse approximately 50% of their water. The U.S. market for water treatment would expand rapidly if Americans convert to drinking water from the tap than consume bottled water. In 2006, Americas bought almost 35 billion bottles of water. Three out of four Americans drink bottled water. It takes 1.5 million barrels of oil per year to produce the plastic bottles for bottled water consumption in America. Many large U.S. cities curtail bottled water usage. Bottled water drinkers are often under the impression that bottled water is safer. The limited freshwater resources and the imbalance of the location of their sources relative to population centers will continue to put pressure on global water supplies driving a need for low-cost disinfection technologies. The development of energy-efficient UV LEDs for water sterilization purposes has the potential to provide a low-cost source of potable water for the world. If efficiency improvements using this technology are maximized, then it is likely that disinfection of water using small UV LED systems could even be powered by solar cells.

An entrepreneur, an essential part of the technology startup, is a person who offers an innovative solution to a frequently unrecognized problem, a person who upsets and disorganizes.

Entrepreneurs innovate. Innovation is the specific instrument of entrepreneurship. The defining characteristic of entrepreneurship is not the size of the company but the act of innovation. Humans are the only species capable of innovation, but these skills are not cumulative, and other animals do not improve their technologies from generation to generation. Entrepreneurs innovate. Innovation is the specific instrument of entrepreneurship. The defining characteristic of entrepreneurship is not the size of the company but the act of innovation. Humans are the only species capable of innovation, but these skills are not cumulative, and other animals do not improve their technologies from generation to generation. Only humans innovate continuously[10]. For entrepreneurship to thrive several specific conditions must be in place, like for a biological cell where everything depends on everything else. But where you need any public-private coordination, we have become handicapped. What is required is rebuilding the infrastructure, so that it is an asset rather than a drag, and reinvesting in research for the industries our grandchildren will found.

Manufacturing, a Strategic Sector

The capital intensive and highly productive American system of manufacturing brings together mind and hand. The U.S. manufacturing sector represents the ninth-largest economy in the world. The average establishment size in manufacturing is three times that of the overall private sector. Manufacturing companies are the source of 70% of R&D performed by U.S. industry, accounting for a total of $206 billion in R&D funding in 2011, and employ 64% of the nation's scientists and engineers. Manufactured goods represent 69% of all exports. Manufacturing is also responsible for 90% of all patents. Workers produce twice as much in manufactured goods per hour roughly every two decades, while it takes three-and-a-half decades for output-per-hour to double in the overall economy. Hourly total compensation in the manufacturing sector is approximately 22% higher than average compensation in service industries. Manufacturing jobs are more likely to offer training. Manufacturers offer these higher wages because of the relatively long tenure and better training of their workforce. Every dollar in manufactured goods generates an additional $1.37 worth of additional economic activity - more than any other economic sector. One study found that each job in manufacturing supported three jobs in the rest of the economy. Ten years ago the U.S. maintained a trade surplus in advanced technology manufactured goods whereas in 2014 this category accounted for an annual $86 billion trade deficit.

Economies where manufacturing skills vanish are less likely to harbor future innovation. The manufacturing innovation ecosystem is a

---

[10]Drucker, Peter - Innovation and Entrepreneurship, 2006

kind of industrial commons; every company shares skills and knowledge that underpin an industry. In the industrial commons participants contribute to the knowledge base and supply-chain rather than just consuming resources. When a company outsources manufacturing it depletes the commons, suppliers close or move, engineers learn different skills, local colleges drop some job-training courses, and the whole ecosystem shrinks. In Germany, Mittelstand companies typically employ fewer than five hundred employees. These employees comprise more than 70% of German workers, and contribute roughly half of that country's GDP. These companies have emerged as successful models in an era of globalization – agile creatures darting between the legs of multinational behemoths. Their blend of high-tech, long-term thinking, and focus on quality has helped German manufacturing through the last recession. This serves as an example of focusing on your traditional strengths in "old-fashioned" industries. Niches that appear tiny can produce large global markets, and companies can preserve high quality jobs in a vast array of industries so long as they are willing to focus and innovate.

New production may not mean a larger manufacturing sector with large numbers of added jobs, but would mean radical change in current technologies and business models. The strong performance of manufacturing in some other advanced industrial countries suggests that manufacturing and blue-collar work are not doomed in high-wage environments. In Germany, where wages and social benefits for manufacturing jobs are higher than they are in the United States, the fraction of the workforce employed in manufacturing is about twice as high as in America. Germany has a manufacturing trade surplus, even in its trade with China. Realizing such possibilities in America will involve a major transformation of aging industrial structures that are often less efficient than the large new plants and industrial complexes of Asia.

It is implausible to think that manufacturing can be outsourced without any damage to our engines of innovation. The great new U.S. companies of the past twenty-five years have been ones with few if any manufacturing capabilities. Many of the vertically-integrated giants like Hewlett-Packard and Texas Instruments have shed their manufacturing, outsourcing much of it to Asian contractors. Will modularity and the separation of innovative activities from manufacturing characterize the new industries of the next decades that promise to transform our economy and society as they have characterized the information technology industry of the recent past? Industries such as biotech, wind and solar power, nanotechnology, aerospace, next-generation automobiles, advanced batteries, and optoelectronics. Challenges in scaling up in such industrial sectors, from laboratories through startups into full production of new products and services, are different from the issues that software or electronics companies face in their transition from product idea to market. Scaling up requires much more capital in these new industries than it does in software. Equally critical is that in today's emerging technology sectors R&D, design, and production are harder to separate than they are in the information technology industry. Only countries that can build powerful links between laboratory research and new manufacturing will be able to derive full benefit from their innovative capabilities.

Quantum Dots for Lighting
Low-cost Manufacturing of Semiconductor Nanocrystals

This startup based in Salt Lake City, Utah is being provided federal funding from September 2014 to August 2016 in the amount of approximately $723,000 for R&D work to develop a low-cost manufacturing method that is urgently needed for production of large-scale and consistently high-quality semiconductor nanocrystal quantum dots. Quantum dots are a class of semiconductor nanocrystals that efficiently absorb and emit light in very precise, tunable wavelengths across the entire visible spectrum based entirely on their size. Given their energy efficiency and infinite color possibilities, their application ranges widely from displays and lighting to solar panels and biomedical applications. The startup manufactures customizable, commercial-scale quantum dots with consistently superior quality creating a unique solution for their displays. The proposed R&D work is based on an innovative proprietary low-temperature method using wet chemical synthesis. Compared to conventional high-temperature synthesis with its scaling limitations, this method can more precisely control the size and shape of products - properties that are necessary for successful incorporation of these products into end-user applications. Current work focuses on demonstrating scaled-up production of larger quantities of high-quality nanocrystals, including heavy metal-free quantum dots.

This effort will help remove key manufacturing barriers that are currently hindering commercialization of semiconductor nanocrystals in diverse market segments worldwide. The unique size- and shape-related properties of these materials make them ideal for light emission applications (including lighting and displays) and light harnessing applications (solar panels). If successful, nanocrystals will be produced in large quantities, inexpensively and uniformly, resulting in a disruptive advance for existing markets and emerging applications. With greater availability and affordability, nanocrystals can be more easily utilized for more energy efficient lighting and displays, improve color quality in displays (laptops, tablets, cameras and mobile devices), increase efficiency of solar panels, and penetrate more widely into advanced applications in medical research, diagnostics and treatment. Emerging applications include the use of semiconductor nanocrystals for biofuel cells, lasers, fiber optics, electronics, security, aviation, and geothermal tracers.

The demand for new, cleaner energy sources for example promises outsize markets for technologies that can be manufactured cheaply enough to compete with fossil fuels. In addition to three-dimensional printing, there are strong new possibilities in bio-fabrication and nanomaterials. The shift in manufacturing is curtailing the development of emerging technologies in areas such as optoelectronics and advanced materials for the automotive industry. The relocation of component manufacturing from the U.S. to East Asia in optoelectronics and to China in composite body parts for automobiles changed the economics of producing the technologies.

The result is that emerging technologies developed in the U.S. were not economically viable to produce in Asian countries because of differences in manufacturing practices. Much of the responsibility for commercializing emerging technologies will fall to small firms, supported by venture capital and government funding, although it is unclear whether those companies can compete in the short-term with larger, multinational firms pursuing older technologies that are currently more cost-effective.

The maker movement is bringing the same technical advances that we saw in media – instant reproduction, endless customization, easy fabrication, and much more widely distributed means of production – to the much larger business of physical things. If professional investors will not risk their money on hardware startups, entrepreneurs can turn to crowd-funding sites for their capital needs, and an ecosystem of hardware service providers to help streamline operations. Low-volume, high-precision, high-mix operations, automated manufacturing, and engineered products requiring technology improvements or innovation are the primary forms of manufacturing returning to America. These are the sorts of products where quality, scrap and inexpensive shipping far outweigh cheap labor.

The Biotechnology Age will do to the physical world what the Information Age has done to data - transition from a world where manufacturing is capital-intensive to one where the cost of manufacturing is on par with the cost of replicating information. Cells will behave both as the factory and the product, where fossil fuels are replaced with genetically-modified (GM) crops and algae, enzyme designers grow plastics straight from bacteria, bacterial cells self-assemble into useful objects, and batteries and solar panels are grown using GM viruses. All this while startup companies in Silicon Valley find rooms full of doctoral-level engineers not seeking cancer cures or sources of safe drinking water for the underdeveloped world but working on schemes to send little digital pictures of emoticons and dragons between adult members of social networks. Given these global trends, the resource constraints, the elements of the American competitive advantages that still persist, and the critical need for fundamental knowledge-based technological innovation in the modern economy, it is imperative to further refine the seeding of whole new industries to continuously create high-wage jobs in the twenty-first century. It is not the false choice of the private sector versus the public sector but the private sector working hand-in-hand with the public sector that would lead to positive outcomes in terms of successful companies and job creation.

The first two chapters of this volume provide a general overview of technology's ubiquitous and persistent presence in modern society, and the importance of innovation in the American economy.

Chapter 3 describes the role of the U.S. federal government as a partner and summarizes the largest five government agencies that invest in discovery science. The following chapter discusses the part played by federal departments in innovation research funding and showcases technologies nurtured via two examples from each of the above-mentioned five federal agencies. Chapter 5 describes the innovation ecosystem, typically made up of entrepreneurs; intellectual property and the patent system; venturesome customers, those early adopters so crucial to the success of startup companies; technology-specific networks; industry partners; manufacturing partners; private investors; research universities; government policies and funding mechanisms; and support provided by local ecosystems.

The next chapter is about demand creation, market entry pathways, go-to-market strategies, and new approaches to gain customer traction. Chapter 7 illustrates probable outcomes and the impact of startup activity through the so-called garage-to-market software startup and the lab-to-manufacture hardware startup pathways, within the confines of the given technology innovation ecosystem. Chapter 8 makes the case for the importance of manufacturing, and why it is critical to continue to make things in America. The final chapter lays out a possible vision for the future and other long-term aspirations, of how a growing stable of high-tech startups, if correctly nurtured can possibly seed whole new industries in the United States.

The author hopes this material would serve to teach, uplift, and inspire potential entrepreneurs to continue to create high-tech businesses in America. This book can benefit startup founders, ninety-nine percent of whom are unable to obtain public funding or private venture capital. This volume can be useful reading for policymakers and think tanks in the U.S. and in other nations, consulting firms, university administrators, industry practitioners, innovation experts, investors, and global organizations such as the United Nations and the World Bank. This endeavor would be considered worthwhile if it is able to assist practitioners, rainmakers, and agents of change become more productive, intellectually and pragmatically more efficient, and frugal too.

Part I

General Considerations

Chapter 1   Society and Technology

It has become appallingly obvious that our technology has exceeded our humanity
– Albert Einstein

The synergistic, cyclical, co-dependent relationship between technology and society occurred from the dawn of humankind, with the invention of simple tools, and continues into modern technologies. Biological humanity and technical civilization, and the human minds that support it, suggest a radically new form of existence, one as different from life as life is from simple chemistry, an increasing mismatch between our stone-age biology and our information-age lives. Our hardwired genetic affinity for life and life processes ensured survival in the past by nurturing our familiarity with nature[11]. We harbor a fondness for made things, in part because we are made, also in part because every technology is our child. Craftsmen have always loved their tools, birthing them in ritual and guarding them from the uninitiated. Machines became a communal experience as the scale of technology outgrew the hand.

Science is knowledge of what things are and how they work. Technology is knowledge of how to do things, make things, and make things do what we want them to. In scientific research, there is a tension between the goals of fundamental understanding and considerations of use[12]. Much technological innovation has proceeded without the stimulus of advances in science. In general the flow is two-way, combining Newton's tradition of understanding the natural world and that of Bacon, of using this understanding to achieve purposive

---

[11]Wilson, Edward - The Future of Life, 2003

[12]Stokes, Donald - Pasteur's Quadrant: Basic Science and Technological Innovation, 1997

needs. Some do not believe any longer that a heavy investment in pure, curiosity-driven basic science will by itself guarantee the technology required to compete in the world economy and meet a full spectrum of societal needs[13]. Technology is essential to science for purposes of measurement, data analysis, computation, communication, and the like. More and more, new instruments and techniques are being developed through technology that makes it possible to advance various lines of scientific research. However, technology does not just provide tools for science; it also may provide motivation and direction for theory and research. The mapping of the locations of the entire set of genes in human DNA has been motivated by the technology of genetic engineering, which both makes such mapping possible and provides a reason for doing so. As technologies become more sophisticated, their links to science become stronger. In fields such as solid-state physics, the ability to make something and the ability to study it are so interdependent that science and technology can scarcely be separated.

The challenge now is to find new avenues to provide for a human society that presently has outstripped the limits of global sustainability. Decarbonizing our technology and creating a sustainable civilization are the overriding goals of our age. The Earth's population has already doubled three times during the past century. We have developed the capacity to dominate most plant and animal species. We have the ability to shape the future rather than merely respond to it. What the future holds for life on Earth, barring some immense natural catastrophe, will be determined largely by the human species. The same intelligence that got us to where we are, improving many aspects of human existence and introducing new risks into the world is also our main resource for survival.

Nature of Technology

Technology, like ritual, morals, and commerce, is an intrinsic part of a cultural system – it both shapes and reflects the system's values. Art, language, and machines are different forms of technology. Technology is an un-hiding, a revealing of an inner reality, this inner reality the immaterial nature of anything manufactured[14]. Technology is a type of thinking, thought expressed. Accepted scientific paradigms elaborate over time, encounter anomalies, and are replaced by new ones. Is this also true for technology[15]? Large systems of technology often behave like primitive organisms - networks, especially electronic networks, exhibit near-biological behavior, and technology systems are

---

[13]Holton, Gerald - Thematic Origins of Scientific Thought: Kepler to Einstein, 1988

[14]Heidegger, Martin - Being and Time, 2008

[15]Kuhn, Thomas - The Structure of Scientific Revolutions, 1996

beginning to mimic natural systems. At its core technology is about ideas and information; both life and technology seem to be based on incorporeal flows of information. A technology, in essence, is a collection of phenomena captured and put to use. Phenomena are the genes of technology, and just as biology programs genes into myriad structures, so does technology program phenomena to myriad uses[16].

Technology now as great a force in our world as nature; our response to both hence should be similar[17]. We are coevolving with our technology and so we have become deeply dependent on it. It is not an extension of our genes but of our minds. We are continuous with the machines we create. We trust in nature, but we hope in technology. Simultaneously we have become imprisoned in a technology framework, in mind-forged manacles[18]. Humans have become an adjunct to or appendages of the machine[19]. Bacon, a founder and champion of modern science, sought not only to highlight the potential of technology to improve human life, but also to foresee some of the social, moral, and political difficulties that confront a society shaped by the great scientific enterprise. He hints at dilemmas that arise with the ability to remake and reconfigure the natural world, so that it might flourish freely without destroying or dehumanizing us[20].

Technology creates our world, our wealth, our economy, our very way of being. Human history through its scientific and technological development begins in prehistoric times and ends with the many accomplishments of the late twentieth-century[21]. To most people, technology has been reduced to computers, consumer goods, and military weapons; we speak of technological progress in terms of random access memories, miracle drugs, and the sleekness of our smartphones. We must restore to technology the conceptual richness and depth it deserves by chronicling the ideas about technology expressed by influential thinkers who not only understood its multifaceted character but who also explored its creative potential, by drawing on an enormous range of literature, art, and architecture to explore what technology has brought to society and culture, and to

---

[16]Arthur, Brian - Nature of Technology: What It Is and How it Evolves, 2009

[17]Kelly, Kevin - What Technology Wants, 2010

[18]Blake, William - The Marriage of Heaven and Hell, 1793

[19]Marx, Karl - The Communist Manifesto, 1948

[20]Bacon, Francis - Of the Proficience and Advancement of Learning, Divine and Human, 1605

[21]McClellan, James and Dorn, Harold - Science and Technology in World History: An Introduction, 2006

---

explain how we might begin to develop technology that works with, not against, ecological systems[22].

The irresistibleness of technology to humans is espoused in the idea that humanity cannot rebuff the temptation of expanding our knowledge and our technological abilities. However, this does not mean that this seeming autonomy of technology is inherent. This perception lingers because humans do not adequately consider the responsibility that is inherent in technological processes[23]. Some insist that technological evolution is essentially beyond the control of individuals or society, that individuals rely on governmental assistance to control the side effects and negative consequences of technology[24]. Technological innovations affect, and are affected by, a society's cultural traditions. From the very beginnings, technology has spurred the development of more elaborate economies. In the modern world, superior technologies, resources, geography, and history give rise to robust economies; and in a well-functioning sound economy, economic excess naturally flows into greater use of technology. Moreover, because technology is such an inseparable part of human society, especially in its economic aspects, funding sources for new technological endeavors are virtually illimitable[25]. The economy is an expression of its technologies; and because an economy arises out of its technologies, it inherits from them self-creation, perpetual openness, and perpetual novelty. The economy therefore arises ultimately out of the phenomena that create technology; it is nature organized to serve our needs.

Role of Technology

The role of technology in society involves social forces that shape technological decisions and the choices that are open to society with respect to its uses. In today's world, technology is a complex social enterprise that includes not only research, design, and crafts, but also finance, manufacturing, management, labor, marketing, and maintenance. The reason for the exponential growth in technology is that the more that is known, the easier it is to discover new knowledge, for knowledge builds on itself. Life now involves a profound tension between the virtues of more technology and the personal necessity of less. We have made newness a virtue and a source of delight, dream

---

[22]Hughes, Thomas - Human-Built World: How to Think about Technology and Culture, 2005

[23]Ellul, Jacques - The Technological Society, 1967

[24]Winner, Langdon - The Whale and the Reactor: A Search for Limits in an Age of High Technology, 1989

[25]Mumford, Lewis - The Myth of the Machine, 1967

of utopias that fantasize better futures rather than recall paradises lost, this pleasure in the new and better, this cultivation of invention and innovation. Progress in science does not map tidily onto progress for humanity. What we have now is more technology, less time, and more stress, less leisure.

A nation's history plays an important role by shaping such things as the base of skills that have been created, prevailing values and norms of behavior, needs, tastes and preferences that will underpin demand patterns, and challenges that have been set or confronted. In the United States, decisions about which technological options will prevail are influenced by many factors, such as consumer acceptance, patent laws, availability of risk capital, the federal budget process, local and national regulations, media attention, economic competition, tax incentives, and scientific discoveries. Culture has a large role to play in a society, and specifically in this context, to ease of adapting to new technologies in a venturesome economy. The economic importance of culture lies in its ability to absorb and harness human talent. Some cultures value education and thought above all else, other societies believe that life should be about struggle and striving, some may look askance at technology, thinking of it as machines, mostly of western origin, and so may be reluctant to perpetuate western dominance and would prefer to continue their nomadic, pastoral or simple life styles; certain peoples show an affinity for numbers, for science and technology, other cultures value elders and old people and indulge in ancestor worship. Do these differing value systems affect the ease of adapting new technology[26]?

Technology itself is incapable of possessing moral or ethical qualities, since technology is tool making. Is it always, never, or contextually right or wrong to invent, develop and implement a specific technological innovation? Ethical questions are sometimes exacerbated by the ways in which technology extends or curtails the power of individuals and how standard ethical questions are changed by the new powers. The ethics of computer security and computer viruses, for example, asks whether the very act of innovation is an ethically right or wrong act. Does a scientist have an ethical obligation to produce or fail to produce a nuclear weapon? What are the ethical questions surrounding the production of technologies that waste or conserve energy and resources, the production of new manufacturing processes that might inhibit employment, or might inflict suffering in the developing world? Bioethics is now largely consumed with questions that have been inflamed by the new life-preserving, cloning, and implantation technologies. In law, the right of privacy is being continually attenuated by the emergence of new forms of surveillance and anonymity. The old ethical questions of privacy and free speech

---

[26]Schlesinger, Arthur - The Disuniting of America: Reflections on a Multicultural Society, 1998

are given new shape and urgency in the internet age. Tracing devices such as radio frequency identification, biometric analysis and identification, and genetic screening, all take old ethical questions and amplify their importance.

Technology will be central to the future of American life and American politics. It will create new political divides and new moral quandaries[27]. The first political consideration pertinent to the role of technology is the proper allocation of resources that society is willing to set aside for the purpose of technology development and adaptation by members of society. This involves investing in people for education, training and workforce skills enhancement, and investments made to improve both the physical and knowledge infrastructure for the purpose of furthering discovery research, pushing the frontiers of science, the feedstock for all technology development that is dependent on knowledge-based innovation. The need to invest both in the physical sciences and life sciences is apparent, not too skewed to one to the detriment of the other, and requires support from the general public in a democratic society. It is far easier to obtain such support for investment in the biological sciences because people directly make the connection to personal health, whereas it is far more difficult to appreciate such levels of investment in the physical sciences, although it helps to remind people that it takes the finest, most sophisticated tools and instruments to push the frontiers in the biological sciences. The convergence of the physical and biological sciences is soon likely along with whole new mind considerations[28], and the talk by futurists of singularities of various kinds[29].

While in the beginning, technological investment involved little more than the time, efforts, and skills of one or a few men, today, such investment may involve the collective labor and skills of many millions. The government is a major contributor to the development of new technology in many ways[30]. Consider the familiar association in America among the military, universities and corporations – a number of government agencies specifically invest billions of dollars in new technology[31]. Technology has frequently been driven by the military,

[27]Hjorth, Linda et al - Technology and Society: Issues for the 21st Century and Beyond, 2007

[28]Wilson, Edward - Consilience: The Unity of Knowledge, 1999

[29]Lanier, Jaron - Who Owns the Future? 2013

[30]Note: The underlying theme for the five instances presented in this chapter is how technologies created by startups funded by the U.S. federal government can provide workable solutions to societal problems such as public health; agricultural worker shortage; food spoilage and child malnutrition; environmental degradation; and universal education.

[31]Kidder, Tracy – The Soul of a New Machine, 1981

with many modern applications developed for the military before they were adapted for civilian use[32]. Cold war defense spending generated

---

Public Health and Entrenched Interests
Flexible Plastic Packaging without Estrogenic Activity

This startup based in Austin, Texas was provided federal funding from September 2011 to December 2014 in the amount of approximately $599,000 for R&D work to develop resins and additives shown to be free of estrogenic activity (EA) to create novel flexible plastic films and products for the preparation and storage of food and beverages that do not leach chemicals having EA. These films and products should remain free of EA when extracted by common solvents and food simulants, and remain EA-free after the stresses of manufacturing, and exposure to common-use stresses such as microwaving, thermal cycling, and ultraviolet light. This research has led to the development and commercialization of food and beverage packaging that are significantly safer, especially for pregnant women, infants, and young children. EA has been strongly linked to early puberty in females, obesity, and increased rates of breast, ovarian, testicular, and prostate cancers. It also leads to reduced sperm counts in males, altered functions of reproductive organs, and altered behaviors and learning rates. Data show that the vast majority of plastic food packaging leaches chemicals with EA, including those advertised as free of Bisphenol-A (BPA). This R&D effort facilitated a comprehensive reduction of risks to public health and reduced environmental impact from chemicals having EA. The company was founded with the mission of creating plastic bottles free of all chemicals with estrogenic activity. Consumers, the popular press, legislators, and nongovernmental organizations are increasingly concerned about the safety of plastics almost all of which release chemicals with estrogenic activity. These chemicals are contained in and leach from many common consumable, packaging, and retail products. While estrogens occur naturally in the body, many scientific studies have reported significant health problems when synthetic chemicals are ingested that mimic or block the actions of natural hormones. These potentially harmful products are widely present in infant feeding products, food storage and preparation bags, cling wraps, pet feed bags, and medical packaging. It is estimated that 20 million baby bottles are sold annually in the United States alone. Consumers are extremely interested in the safest baby bottles available, as evidenced by the growing market penetration of glass and stainless steel bottles. The company's business model is based on the fact that regulation is not a prerequisite for rapid market change but the threat of later regulation/litigation is helpful.

---

an industrial commons that spilled over into commercial production. The military-industrial complex invented small transistors eventually used in radios and satellites and wristwatches; the Department of Defense researched and produced hard plastics, optical fibers, lasers, computers, jet engines and aircraft frames, precision gauges and sensing devices. These led to graphite tennis racquets, remote-controlled televisions, microwave ovens and cellphones. The Pentagon gave birth to the nation's first computers, new fighter jets and engines morphed into commercial jet aircraft; work at the Defense Advanced Research Projects Agency led to the birth of the Internet. However, this

---

[32]Reich, Robert - Aftershock: The Next Economy and America's Future, 2010

has always been a two-way flow, with industry often developing and adopting a technology only later adopted by the military. Research and development is now one of the smallest areas of investments made by corporations toward new and innovative technology.

Eliminate Back-breaking Work
Machine Learning Applied to Robust Crop and Weed Detection

This startup based in Mountain View, California is being provided federal funding from April 2013 to February 2016 in the amount of approximately $1.0 million for R&D work to further develop a novel computer vision-based plant identification system for agricultural weed control. This system will provide a cost-competitive alternative to the $20 billion global chemical herbicide market. Existing computer vision-based approaches can segment a 'splotch' of green vegetation from a brown background but are unable to provide the selectivity and precision necessary for mechanized, automated weeding. This startup's objective is to create software algorithms based on a hierarchical classifier that matches the capability of the human eye and brain to quickly and reliably classify plants into crops and weeds in real-time. This classifier will utilize a novel approach to visual object identification using a field-customized support vector machine that uses point-of-interest rather than shape-based methods. The research will result in the creation of an algorithm integrated into a real-time, automated weeding machine and plant identification system, and advance the fields of computer vision and machine learning. The impact of this R&D work is the development of an alternative to chemical intensive agricultural weed control. The system will offer conventional farmers an effective and chemical-free method to eliminate weeds, and offer organic farmers the first truly precise organic weed control method. The market for weed control in food production in America is estimated to be $4 billion. U.S. farmers apply over 250 million pounds of herbicide annually on corn and soybeans alone, with many unintended and detrimental side effects. Chemical concentrations in rivers, lakes and groundwater are rising, and the prevalence of herbicide resistant weeds is growing exponentially. An alternative to these chemicals limits society's exposure while protecting environmental integrity.

The company's mission is to provide advanced technology for better agriculture. The startup's approach utilizes computer vision and robotics to build a future in which "every plant counts", where the needs of each plant are precisely measured and delivered to significantly reduce chemical use. The startup developed LettuceBot, a precision thinning system that gives the farmer more control than ever before, through cutting-edge robotics and machine learning algorithms that make plant-by-plant decisions to increase yield, ultimately earning more value per acre. This startup's first commercial technology, precision thinning, serves some of the largest lettuce growers in the country, by visually characterizing each plant; calculating which to "keep" to optimize yield; and precisely eliminating unwanted plants, all in rugged agricultural conditions, automatically, and in real time. Their lettuce thinning service now operates in California's Salinas, Central and Imperial Valleys as well as in and around Yuma, Arizona.

The second political consideration pertinent to the role of technology pertains to policy matters that involve strategy that often seeks bipartisan support. It requires a version of industrial policy, a dreaded phrase in America these days, as if this has not all along been a part of our thinking in society and government. The time horizons for crucial investments, steady, careful and well-thought through solutions over the long term, unlike Wall Street's, and therefore corporate America's obsession with the next quarter, investments into solutions that target, attack and solve society's grand challenges. The need for tax and credit policies related to technology initiatives, reduction of waste and abuse in government or public spending, specific policy measures crafted for a long-term innovation agenda, and support for certain critical industries, to form the next set of whole new industries for future generations through effective public-private partnerships. A strong urban policy is as important to our nation's future as a strong innovation policy, and upgrading of the physical and cyber infrastructure, for example, broadband access to rural and under-served urban areas[33]. Investment and innovation in schooling and education - different kinds of school systems, changes in tenure policy, changes in union structures, improved and higher pay-scales for teachers, merit-based performance benefits and recognition, higher status for teachers in society, and the involvement of parents in their children's education across all sections of society not only the middle and upper classes.

Technology Evolution

The evolution of technology, like an ever-ascending staircase, with one novel development set atop another, has led incrementally and inevitably to the benefits of modern life[34]. This evolution is inevitable, we cannot stop it but the character of each technology is up to us. If human beings survive, we are destined to evolve further with our own active participation in directing the process, towards a convergence between humans and machines, to not only understand the blueprints of life-forms, but also to change their designs and functions. It is a wondrous coincidence that human knowledge and experience can be expressed in binary digital terms, this science movement now driven by the digitization of life and human beings, a frightening prospect to some, with overlapping and still accelerating revolutions in genetics, proteomics, regenerative medicine, neuroscience, and bioinformatics. The first technological revolution in modern biology started a half-century ago when Watson and Crick discovered the double-helical structure of deoxyribonucleic acid (DNA) by witchcraft-like techniques of the biochemist[35]. This established molecular and cellular biology, the basis of the biotech industry.

---

[33]Florida, Richard - Flight of the Creative Class: The New Global Competition for Talent, 2007

[34]Gertner, Jon – The Idea Factory: Bell Labs and the Great Age of American Innovation, 2013

[35]Watson, James - The Double Helix: A Personal Account of the Discovery of the Structure of DNA, 2001

Sequencing of the human genome nearly a decade ago set off a second revolution which has started to illuminate the origins of disease. A third revolution underway now, when Venter and Smith instead of giving a creature its living essence, they made that essence with a thousand-gene piece of DNA from off-the-shelf lab chemicals[36]. The result is the first creature since the beginning of creatures that has no ancestor. When the first of these artificial creatures showed it could reproduce on its own, the age of artificial life began although evolution by artificial selection likely to prove almost as wasteful as the kind by natural selection.

The bigger tasks than that of the human genome project is the mapping of all the proteins that are expressed by genes, complete the first larger-scale maps of neural wiring, and sequence the genome of every human being alive. By transferring genes from one species to another and by creating novel DNA strands of their own design, scientists can insert them into life-forms to transform and commandeer them to do exactly what they want them to do, to even program the replication of the inserted DNA strands. Like viruses, these DNA strands are not technically "alive" because they cannot replicate themselves, but also like viruses, they can take control of living cells and program behaviors, including the production of custom chemicals that have commercial value. Ancient boundaries are being crossed, boundaries that separate one species from another, the divide between people and animals, the distinction between living things and man-made machinery, by introducing human genes into other animals, by creating hybrid creatures mixing genes of bees and baboons, by cloning humans, and by surgically embedding silicon computer chips into human brains, providing a genetic menu of selectable traits for parents who wish to design their own children.

Technology will soon enable us to print-on-demand vaccines and pharmaceuticals from basic chemicals and for people to grow their own replacement organs via three-dimensional (3D) printing. We will have nanofactories producing proteins while inside the human body; nanoscale machines for insertion into humans with active devices the size of living cells that can co-exist with human tissue, and nanobots to monitor changes in the bloodstream and in vital organs. Intelligent prosthetics that will allow a person who has lost both arms to play the piano again and robotic suits that interact with damaged nervous systems to enable people to walk will be available. Synthetic brain subsystems will be deployed for paraplegics to control advanced prosthetic arms and legs with their brains as if they were using their own natural limbs by connecting the artificial limbs to neural implants. Such subsystems would be used to enhance humans, to make a human sperm-making biological machine, to convert functional sperm

---

[36]Venter, Craig - Life at the Speed of Light: From the Double Helix to the Dawn of Digital Life, 2013

cells from stem cells, and to protect telomeres to retard the aging process. DNA's information storage capacity is incredible – recently a research team encoded a fifty-thousand word book into strands of DNA and read it back with no errors. Programmable materials that could

---

Technology Looking for Problem
Thermal Energy Storage to Prevent Food Spoilage

This startup based in Cambridge, Massachusetts was granted federal funding from April 2013 to September 2015 in the amount of approximately $591,000 for R&D work to commercialize a new type of industrial refrigeration system. The innovation lies in a novel thermal battery pack concept developed by the company that converts and stores electrical energy in the form of thermal energy through a process of freezing and melting a material. The battery needs just five hours of grid electricity to charge and, once charged, provides constant cooling power on demand. The impact is to introduce clean energy technologies in refrigeration applications, eliminating costly spoilage of food in developing economies that currently rely on fossil-based fuels for their continued growth. In such countries, billions of dollars of perishable foods are wasted annually because of inadequate cold-chain supply networks. A major obstacle in setting up a cold-chain network is the lack of reliable grid electricity to run refrigeration systems in villages and farming areas. In such conditions, diesel generators are often used as backup, a non-ideal solution with high cost and environmental impact. This startup's thermal battery eliminates diesel generators and makes village refrigeration systems economically viable. This could revolutionize dairy industries around the world, especially in India, where nearly 400 million people depend on milk for their daily protein.

The company started with a solar concentrator designed for a developing market. This was a classic case of a technology looking for a market – using the sun to produce hot water and electricity but for what market? The two founders met with the Managing Director of Bangalore Dairy - this company was not too interested in the solar concentrator. Instead they told the founders about their problem – milk spoilage. In 2011 the startup delivered its second-generation solar-powered refrigeration system to chill 500 liters of milk per day, to India's largest private dairy owner. This customer upon inspecting the system said it was too expensive, took up too much space, was not cooling milk fast enough, and that it was impossible to have such a large, complicated unit in dozens of villages and towns. After much deliberation the team wondered if they would have to cast aside solar power. Everything they had worked for was to use solar power but it was turning out to be the wrong solution. They resolved to build another prototype – the same problem persisted – how to chill milk when there is no electricity supply? Suddenly it dawned on them that the energy storage device was the key. When they had previously relied on solar power, a large tank of cold water was used as an energy storage mechanism for night-time operation or as a backup during cloudy days. The critical insight was that the energy storage device was the element that can operate together with grid electricity like a battery to fill the gaps in grid power. Their thinking was obscured by the need to use solar power and they had missed the solution, the true innovation.

---

form and reform themselves into different useful shapes on command and growing plants that respired hydrogen to be harvested as a cheap source of fuel are distinct near-future possibilities.

The next wave of innovation in information and communications technologies will most likely involve cognitive radios that allow for more efficient sharing of the electromagnetic spectrum, quantum computing, robust programming of parallel computers, cyber-physical systems, secure computers and networks, data-intensive supercomputers, and nanoelectronics that will perhaps enable the continuation of Moore's Law for decades to come. An exascale supercomputer capable of a million trillion calculations per second will dramatically increase our ability to understand the world around us through simulation and slash the time needed to design complex products such as therapeutics, advanced materials, highly-efficient cognitive automobiles, and aircraft. Nanotechnology promises to

Make Education More Accessible
Simulated Patients for Medical Training

This startup based in Sunnyvale, California is being provided federal funding from September 2013 to February 2018 in the amount of approximately $1.25 million for R&D work to enhance the technical understanding of how to create rich, high-performance, entirely cloud-based applications that are accessed anytime, anywhere, on any device. The layered web architecture eliminates the need for a local application or proprietary browser plug-ins like Java or Flash. The framework will provide complex interface components and powerful functionality in composition, styling, dynamic behavior, and data manipulation. Benefits include smooth, seamless local application-like user experience independent of user platform; lower support costs as all business logic is deployed in the cloud; improved security as the server validates all actions so if client data is tampered with, the server can detect it and react accordingly; simpler yet more powerful application programming model; the ability to deliver complex computation and storage-intensive functionality to mobile platforms; and integration of real-time collaboration without local and proprietary installation requirements.

This simulated patient service will enable medical educators to accelerate development of students' clinical competencies in an affordable and scalable manner. This will help address the widely anticipated shortage of qualified U.S. providers driven by the aging of baby boomers and an expected 15% increase in demand for healthcare services due to coverage of the previously uninsured. Shortages in developing countries are even greater. The service will also help ensure the quality of medical education and lower the cost of healthcare as mastery of critical diagnostic reasoning is the basis for efficient delivery of care and is required for optimal outcomes, cost control, and reduction of hospital readmissions. If used in testing, certification, continuing medical education, and protocol adherence, it could ensure the competency of recruits, help busy clinicians keep up with advances in medical knowledge, and identify those in need of additional or remedial training.

transform multiple industries, for instance, the capturing and storing of clean energy, next-generation computer chips, and enabling all-new approaches to a wide range of manufacturing activities and to a range of consumer goods from healthcare and food products to textiles, automotive composites and industrial coatings. Carbon fiber, extremely strong and very light, could become the material of choice to mass-produce electric cars.

Can there be a source for industry away from electricity; perhaps the use of radio waves to generate and distribute electricity? Low-cost, efficient energy storage is the key to making renewable energy more practical. Power saved through conservation or efficiency measures is arguably the best way to meet rising demand for power, both for business and the environment. High-capacity, high-efficiency, smart electricity grids are essential in order to use intermittent sources of electricity like those produced by windmills and solar panels. Solid-state lighting could decrease global electricity consumption ten percent by 2025. Light emitting diodes can light an entire rural village with less energy than that used by a single conventional hundred-watt incandescent bulb. Developing solar cells as cheap as paint, fitting the contents of the library of congress on a device the size of a sugar cube, educational software as compelling as the best video game and as effective as a personal tutor, online courses that improve the more students use them, and a rich, interactive digital library at the fingertips of every child are all near-term possibilities.

Turing was unable to offer an overwhelming argument that a machine can be made to think[37]. By 2020, computers will achieve the memory capacity and computing speed of the human brain, will be able to read on their own, and understand and model what they have read. They would gather knowledge on their own, effectively becoming self-programming thinking machines, and share knowledge with each other which machines can do far more easily than their human creators. Computers learn you as opposed to you having to learn them; your computer will be with you from the moment you are born and grow up, get to know you over time, know what your habits are, know what you want to see, know how you take in information, and know when you want information. Once it achieves a human level of ability in understanding abstract concepts, recognizing patterns, and other attributes of human intelligence, it will be able to apply this ability to a knowledge base of all human-acquired, and machine acquired knowledge.

Artificial reality may be an even more profound concept than artificial intelligence. Thought might then artificially perfect the thinking instrument itself[38]. Machine competence will rival our marvelous ability to place our ideas in a broad diversity of contexts, our ability to respond to abstractions and subtleties. Once robots are as complex as the human brain, and can match the human brain in subtlety and complexity of thought, are we to consider them conscious? Will such robots of superior intelligence and acuity make their own decisions, or will they just follow a program, albeit a very complex one[39]? Adaptive

---

[37]Turing, Alan – Computing Machinery and Intelligence, 1950

[38]Chardin, Pierre Teilhard de – The Phenomenon of Man, 2008

[39]Kurzweil, Ray - The Age of Spiritual Machines: When Computers Exceed Human Intelligence, 1999

---

second-generation robots will find jobs everywhere and may become the largest industry on Earth. As machines increasingly design, diagnose, and repair themselves, automated industry will grow away from Earth, and old Earth will become insignificant on the ever-grander scale of Earth-spawned activity. Around 2030, third-generation robots will develop a kind of consciousness, be able to imitate, adapt, and create simple programs of their own. Social models of third- and fourth-

---

Reduce Kidney Disease Infections
Novel Peritoneal Dialysis Catheter

This startup based in Mountain View, California is being provided federal funding from August 2012 to October 2016 in the amount of approximately $1.1 million for R&D work to improve the quality of life of dialysis patients through the development of a new disinfection process of the peritoneal dialysis (PD) catheter connection to reduce infections related to PD. The prevalence of PD among dialysis patients is increasing and infection rate has become a critical target for improvement. Since PD is self-administered, patient adherence to the set-up protocol is vital to ensure that sterility is observed. Unfortunately, this is not always the case - infection remains a pressing concern. A high number of PD patients experience severe peritonitis (infection) due to a lack of compliance with rigorous disinfection throughout the treatment process. Peritonitis is the bacterial or fungal infection of the peritoneum, resulting in the inflammation, or irritation, of a thin layer of tissue that lines the inside of the abdomen. Peritonitis requires immediate medical attention to fight the infection prior to invading the entire body.

The objective of this R&D effort is to circumvent the need for complete patient compliance, and introduce a device that will decontaminate the interfaces of PD catheters that are at most risk of contamination. The R&D work focuses on enhancing the device design in an effort to achieve a multifold logarithmic reduction on multiple pathogens, including bacterial and fungal. It will also optimize product design, from both performance and usability standpoints. When implemented in the clinic, the disinfection device will allow peritoneal dialysis patients to safely receive the full benefit associated with this mode of treatment. This R&D effort will help develop a new standard of care for all patients suffering from catheter-related infections. The total addressable market for dialysis, central venous, and peripheral venous lines is $13 billion. Catheter-related infections still presents a dangerous health hazard for many markets and as of today no disinfectant has emerged to successfully address this issue. The "no pay" rule implemented in recent years prevents hospitals from being reimbursed for infection-related cases with catheters. The disinfection device will enable the decontamination of catheters and the prevention of luminal infections, thereby greatly reducing the risk of secondary infection. For peritoneal dialysis specifically, this technology allows for home dialysis to be a more attractive option compared to hemodialysis, by adding safety, and reducing patient morbidity and mortality.

---

generation robots represent beliefs about intentions and feelings. When a robot analyzes its own behavior using these models, it creates beliefs about its own feelings. Does it then have genuine feelings, does it only believe it has them, or is it only behaving as if it believes it has them? Should we be afraid that robots become self-aware[40]?

---

[40]Nielson, Donald - A Heritage of Innovation: SRI's First Half Century, 2006

Eventually thoughtful scientist robots may develop powerful models that elucidate the tangled interactions of our kind of consciousness, allowing them to reliably create human-sized minds with particular properties. By then their own minds, greatly expanded and improved by trial and error, will be far more elaborately organized than ours and in operation probably as bewildering to them as ours are to us. They may have dreams and nightmares, and may find some kinds of anger useful in interactions with irresponsible humans or robots. A growing ecology of artificial life will form that will eventually surpass the existing biosphere in diversity[41].

Technology and Its Discontents

Technology is natural for it comes from a natural source, the human mind. In its deepest essence, technology is natural, profoundly natural, but it does not feel natural. As humans we are in the deepest recesses of our being attuned to nature. We trust nature; as humans our greatest hope lies in technology but our greatest trust lies in nature. Technology is part of the deeper order of things but in our conscience we make a distinction between it enslaving versus extending our nature. Technology seems now to be moving from using nature to intervening directly with nature leading to a clash between what technology offers and what we feel comfortable with. Technologies that were supposed to liberate us have invaded our lives. Technology at this stage in human history weighs on us, weighs on our concerns, and has also brought upon us a profound unease. The promises and threats of technology for the future of society and the environment, the sociology of invention, the adoption and diffusion of technology, the ideas of cost-push and demand-pull, the role of institutions and of learned societies, the history of technology - in this territory Heidegger[42] has left traces and Schumpeter's footprints are everywhere[43].

Powerful currents of technological change, the digitization of work, robots replacing humans, and economic determinism will result in a hyper-connected, tightly integrated, highly interactive, and technologically revolutionized economy. The tendency of the industrial age is to fragment work into tasks so trivial that they are fit to be performed only by the equivalent of slave labor[44]. A large percentage of technology workers in America are not content with their jobs.

---

[41]Moravec, Hans – Robot: Mere Machine to Transcendent Mind, 1999

[42]Safranski, Rüdiger and Osers, Ewald - Martin Heidegger: Between Good and Evil,1999

[43]Schumpeter, Joseph - Essays: On Entrepreneurs, Innovations, Business Cycles, and the Evolution of Capitalism, 1989

[44]Ruskin, John – The Stones of Venice: The Sea-Stories, 1853

Among the reasons cited are the nature of the jobs themselves and the restrictive ways in which they are managed; among the terms used to describe their malaise are declining technical challenge, inappropriate utilization, limited freedom of action, and tight control of working patterns. Machine learning will threaten white-collar, knowledge-worker jobs just as machines, automation and assembly lines destroyed factory jobs in the nineteenth and twentieth centuries. Will the progressive substitution of intelligent machines for human labor result in increased structural unemployment, or will we find ways to create new jobs and adequately compensate those filling them? The shift in the relative value of technology to labor, the hemorrhaging of jobs to intelligent machines, and where finance is preeminent in the economy, will lead to a secular, systemic shift to an economy with far fewer jobs relative to production. This would represent a larger cause of declining incomes, and thus declining consumption and demand.

There is no shortage of work to be done, but the dominance of corporations and market encroachment into the ideals of democracy has taken a toll on the initiative and will necessary to structure new employment opportunities in the creation of public goods in fields like education, environmental remediation, physical and mental health, family services, and community building. Will the social compacts in developed nations survive the simultaneous effects of demographic changes that are placing heavier per capita burdens on those in the workforce even as jobs and incomes are lost to the combination of outsourcing and robot deployment? Will new models for restoring income support and healthcare to the growing population of older people be created to replace the twentieth-century model?

There are often constraints on the openness of science and engineering that are relevant to technological innovation because of the economic value of technology. No scientific or technological knowledge is likely to remain secret for long. Secrecy most often provides only an advantage in terms of time, a head start, not absolute control of knowledge. Commercial advantage is not the only motivation for secrecy and control of it could be technology that has potential military applications. The connections between science and technology are so close in some fields that secrecy inevitably begins to restrict some of the free flow of information in science as well. Most technological innovations spread or disappear on the basis of free-market forces, on the basis of how people and companies respond to such innovations. Occasionally, however, the use of some technology becomes an issue subject to public debate and possibly formal regulation. The long-term interests of society are best served, therefore, by having processes for ensuring that key questions concerning proposals to curtail or introduce technology are raised and that as much relevant knowledge as possible is brought to bear on them.

Is modern technology unable to provide both economic prosperity and a clean environment? Pollution is a serious problem in a technologically advanced society, from acid rain to Chernobyl, Bhopal, and Fukushima. The increase and proliferation in transportation technology has brought congestion in megacities. New dangers exist

---

Ameliorate Environmental Degradation
Waste Methane Gas to Biodegradable Biopolymer

This startup based in Palo Alto, California is being provided federal funding from February 2013 to July 2016 in the amount of approximately $824,000 for R&D work to use waste methane gas as a feedstock to produce pellets of polyhydroxyalkanoate, a valuable polymer that is converted into a variety of high-margin or high-volume, eco-friendly plastic products such as children's toys, electronic casings, water bottles, and food packaging containers. The current plastics market is dominated by petroleum-derived, non-biodegradable, energy-intensive plastics, which often persist in the environment upon disposal. Alternative plastics are derived from rapidly renewable bio-based resources and consumed by microbes when no longer needed making them biodegradable. Unfortunately, these alternative plastics are often costly to produce and their manufacturing process requires significant amounts of energy. The company has a novel, patented, energy-efficient method to produce a biodegradable, bio-based polymer at a price competitive with petrochemical-based polymers. This R&D effort will scale the startup's process to produce samples for customers to test while addressing associated challenges. The key goals are to optimize the production process and to verify that customers can make the product on existing manufacturing equipment. Widespread production of low-cost bioplastics from waste biogas and the eventual displacement of petroleum-based plastics will result. Bioplastics have the potential to capture an increasing fraction of the plastics market, thereby giving consumers the choice to purchase affordable, environmentally friendly products. When products made from this startup's bioplastic are disposed in modern wastewater treatment plants or landfills they biodegrade anaerobically, that is without oxygen, to methane. This methane can be cycled back to re-enter the process as feedstock to produce more polyhydroxyalkanoate. This closes the life cycle, creating a cradle-to-cradle system. This use of biogas will provide a strong economic incentive for facilities to capture their methane, rather than releasing or flaring it. This in turn will reduce greenhouse-gas emissions and its corresponding impact on global warming.

---

as a consequence of new forms of technology, such as the first generation of nuclear reactors. New forms of entertainment, such as video games and internet access could have possible deleterious effects on areas such as academic performance, and the social separation of singular human interaction. There is increased probability of diseases and disorders, such as obesity and diabetes, becoming epidemic. The solution to the climate crisis and ocean acidification, global warming's evil twin, will be the central organizing principle of global civilization. The Earth's ecosystem will collapse if India and China follow the same development paths of the West with its over-dependence on fossil fuels, rampant consumerism, and its focus on leisure and entertainment. The pressing needs of billions of people, practically four-fifths of the world's population are mired in poverty, degradation and hopelessness, leads to frustration and violence against the better-off. Policies and investments to focus on providing basic needs to all people on Earth, using technology to do so, are urgent. Food, clean air, drinking water, nutrition and health, sanitation, basic infrastructure, education – these are the grand challenges that require radical breakthroughs for the benefit of humanity.

The power of self-replication is already found in genomics, robotics, nanotechnology, and informatics. Technology itself will create new technology with then an important requirement that it needs to know when to stop replicating. We are moving towards smart systems that are self-configuring, self-optimizing, cognitive, self-assembling, self-healing and self-protecting. Producing synthetic genomes and synthetic DNA will lead to the impending creation of completely new forms of artificial life capable of self-replication, and will produce a new wave of organisms, an artificially provoked neo-life. Fetal genomic screening and trait selection, deciding who deserves to be born, will result in an objectification of children as commodities thus creating a techno-eugenic rat race. This synthetic biology may supplant fifteen percent of the global chemical industry within the next few years. Yet by dint of economic imperative as well as the sheer volume of scientific and commercial activity underway around the world, it is functionally unstoppable, a juggernaut already beyond the reach of governance.

Will the emergent potential for altering the fabric of life and the genetic design of human beings be accompanied by the emergence of wisdom sufficient for the far-reaching decisions that will soon confront us, or will these technologies be widely dispersed without adequate consideration of the full spectrum of consequences they could entail? It poses the most profound challenge to human society's oversight of technology in human history, carrying with it significant economic, legal, security, and ethical implications that extend far beyond the safety and capabilities of the technologies themselves.

Chapter 2   Innovation Economy

I think there is a world market for maybe five computers
— Thomas Watson, Chairman, IBM, 1943

Even if some maintain that economics is ideology masquerading as science, the distinctive character of the modern economy involving uncertainty, ambiguity, and diversity of beliefs provides a revealing perspective on the interaction between innovation and globalization, and its meaning for our long-run prosperity[45]. The core strength of our culture is its dynamic adaptability to accommodate new people and new economic realities. One key factor to America's economic growth is a democratic society that is open to new ideas and rewards merit. This attracts the world's best and has allowed it to dominate the global competition for talent, and to thus harness the creative energies of its own citizens and from people around the world.

The United States retains profound strengths including a system of higher education and an entrepreneurial community that makes it the world's most powerful engine of innovation which in turn drives productivity growth. Sophisticated markets and institutions foster intense competition that help companies discover new paths to higher productivity. This allows more productive firms and institutions drive out less productive ones making the economy dynamic and resilient. Technological advances may reduce overall employment, causing mass unemployment as workers are displaced by machines. At the same time, technological improvements create new products and services, shifting workers from older to newer activities[46]. Higher productivity raises incomes, increasing demand for labor throughout the economy. This is because when labor becomes more efficient, capital is freed up for other, more productive uses. A company that lowers its labor costs through increasing productivity may have extra capital to spend on equipment. That equipment must be purchased and the manufacture of such equipment requires labor. Hence, some

---

[45]Phelps, Edmund – Mass Flourishing: How Grassroots Innovation Created Jobs, Challenge, and Change, 2013

[46]Autor, David and Katz, Lawrence - Grand Challenges in the Study of Employment and Technological Change, 2010

economists maintain that in the long run technological progress affects the composition of jobs not the number of jobs.

Over the last century, enlightened public policy sustained by durable political coalitions allowed America to build large, stable innovation systems that drove extraordinary technological transformations in comparably vast regions of the economy, including agriculture, telecommunications, defense, and health[47]. Any strategy to secure our prosperity recognizes the importance of innovation for sustainable growth and quality jobs. Entrepreneurial capitalism where entrepreneurs provide radical ideas that meet the test of the marketplace, play a central role in this economic system. Technological advance or increase in total factor productivity is the most important source of growth[48]. To transform new ideas into higher productivity and wealth the market also needs the support of economic pillars such as world-class universities, private property, venture capital, and the stock market. The university system performs sixty percent of basic research done in America. Scientific discovery in general is a public good that is underprovided by market forces because scientific knowledge is a non-rival good that can be used by anybody without lessening its availability for everybody else. Since this should remain publicly available, nonmarket means must be used to support the financial investment of resources into such activity.

Private firms rarely sponsor fundamental research at universities, or work to increase the pool of human resources, or invest in building up the supply-chain. If core infrastructure is left to the private market, there will tend to be under-provisioning, monopoly prices, and exclusion of the poor[49]. At the same time, the impetus to innovate, skills to do so, and signals that guide its directions must come largely from the private sector. Government efforts are best spent in indirect ways such as stimulating creation of more advanced factors of productivity, improving the quality of domestic demand, encouraging new business formation, and preserving domestic rivalry. Business leaders can and must play a far more proactive role in transforming competition and investing in local communities rather than being passive victims of public policy or hostages to misguided shareholders. Many large firms invest abroad, and minimize or altogether avoid paying domestic taxes. This migration of business activity out of the U.S. and accompanying loss of revenue makes it harder for government to invest adequately in public goods making America a less attractive place to do business. Critical technologies

---

[47]Lester, Richard and Hart, David - Unlocking Energy Innovation: How America Can Build a Low-Cost, Low-Carbon Energy System, 2011

[48]Baumol, William et al – Good Capitalism, Bad Capitalism, and the Economics of Growth and Prosperity, 2009

[49]Sachs, Jeffrey - Common Wealth: Economics for a Crowded Planet, 2009

such as biotechnology, life sciences, machine tools, automobiles, optoelectronics, information and communications, shipbuilding, electronics, aerospace, and nuclear technology have been at the core of America's economic growth. We are exiting businesses that are strategic to our nation's industrial base, to our ability to compete, and to our long-term future by pandering to shareholders. As the cost of building infrastructure rises exponentially, the price of reentry if we lose this infrastructure becomes overwhelming[50]. If a country keeps competitors out of an industry whose capital investment requirements are steadily rising at some point market entry will be prohibitively expensive for U.S. companies without government help. This strategy is a real threat to long-term U.S. technological competitiveness because of the willingness of other governments and industries to use it[51].

The U.S. innovation system has been weak at deploying and implementing new technologies where there is a consequential need for coordination and public-private cooperation. There is also a significantly reduced demand for capital to fund real wealth-creating activities in America such as capital to finance factories, equipment needed to modernize and expand, new creative and fast-growing startups, and to invest in R&D to create the next generation of products and services[52]. In the U.S. new ventures provide over ten million jobs annually, which accounts for about eight percent of all American jobs. Many large corporations having cut back their R&D expenditures, a growing share of innovation is therefore now occurring in publicly-funded laboratories and in small- and medium-sized firms, many of which are supported with government research dollars[53].

Innovation Economics

Economics is concerned with scarcity -- limited resources in the face of limited desires. If Adam Smith is the patron saint of classical economics[54] and John Keynes of Keynesian economics[55], it is Joseph

---

[50]Elkus, Richard - Winner Take All: How Competitiveness Shapes the Fate of Nations, 2008

[51]Prestowitz, Clyde – Three Billion New Capitalists: The Great Shift of Wealth and Power to the East, 2005

[52]Atkinson, Robert and Ezell, Stephen – Innovation Economics: The Race for Global Advantage, 2012

[53]Landes, David - The Wealth and Poverty of Nations: Why are Some So Rich and Some So Poor, 1999

[54]Smith, Adam – An Inquiry into the Nature and Causes of the Wealth of Nations, 1759

[55]Keynes, John – The General Theory of Employment, Interest, and Money, 1965

Schumpeter who plays such a role for innovation economics[56]. Economic resources for all social activities of human beings are common: capital withheld from current consumption and allocated instead to future expectations; physical resources such as land, infrastructure, and extractive materials; and labor, management, and time. Empirical evidence worldwide shows the strong link between technological innovation and economic performance. Innovation economists believe that what primarily drives economic growth in today's knowledge-based economy is not capital accumulation, but innovative capacity spurred by appropriable knowledge and technological externalities. Here, economic growth is the end-product of knowledge. Innovation economics reformulates conventional economics theory so that knowledge, technology, entrepreneurship, and innovation are positioned at the center of the model rather than seen as independent forces that are largely unaffected by policy[57]. It is based on two fundamental tenets: that the central goal of economic policy should be to spur higher productivity through greater innovation, and that markets relying on input resources and price signals alone will not always be as effective in spurring higher productivity, and thereby economic growth. An innovation economy depends on intellectual property law, patent procedures, tax codes, export controls, immigration regulations, and venturesome consumption – the willingness and ability of business and consumers to effectively use products and technologies derived from scientific research[58].

Well over half of the growth in U.S. output per hour during the first half of the twentieth century could be attributed to advancements in knowledge, particularly technology[59]. While the neoclassical tradition identified investment in physical capital as the driving factor the endogenous growth theory put the emphasis on the process of knowledge accumulation, and hence the creation of knowledge capital[60]. The concept of social capital is a further extension because it identifies a social component to those factors shaping economic growth and prosperity[61]. This refers to connections among individuals - social networks and the norms of reciprocity and trustworthiness that arise from them. A large and compelling literature has emerged that

---

[56]Schumpeter, Joseph - Capitalism, Socialism and Democracy, 1942

[57]Arthur, Brian - The Nature of Technology: What It Is and How It Evolves, 2009

[58]Bhide, Amar - The Venturesome Economy: How Innovation Sustains Prosperity in a More Connected World, 2008

[59]Solow, Robert – A Contribution to the Theory of Economic Growth, 1956

[60]Romer, Paul - Endogenous Technological Change, Journal of Political Economy, 1990

[61]Coleman, James – Social Capital in the Creation of Human Capital, 1988

links social capital to entrepreneurship[62]. These studies indicate that entrepreneurial activity is enhanced where investments in social capital are greater[63]. Profiting from technological innovation is a key strategic challenge in technology-intensive industries because it requires not only scientific and engineering expertise but also an understanding of how business and legal factors facilitate commercialization. Sustained competitive advantage can be accomplished only through continued innovation. Entrepreneurship capital is defined as those factors influencing and shaping an economy's milieu of agents in such a way as to be conducive to the creation of new firms. These concepts of social and entrepreneurship capital suggest that these too are important factors shaping economic performance[64]. The empirical evidence suggests that there is indeed a positive link between entrepreneurship capital and regional economic performance.

Economies benefit not merely from the invention of new products and services but from their diffusion. Infrastructure provides a reliable platform for economic development. In the past, goods were the basis of the economy and the key infrastructures focused on linking producers with markets. Knowledge is now replacing goods as the focus and requires a new kind of infrastructure that links information and expertise, and enables their use in collaborative problem-solving. Cyber-infrastructure will provide a platform for locating and accessing information, specialized resources, collaborators, and experts independent of physical location. Economic value having migrated from goods to knowledge will now increasingly move to data and the algorithms used to analyze them. In fact data and the knowledge extracted from them may even be on their way to becoming a factor of production in their own right just like land, labor, technology, and capital. In this world the ability to imagine and generate new ideas with speed and to implement them through global collaboration is the most important competitive advantage.

Productivity growth is the closest economics gets to a magic elixir, especially for ageing advanced economies. As we leverage our sophisticated new technologies to develop novel products and services, just about every industry will be transformed and achieve major productivity growth. We may be experiencing a fundamental reshaping of the economy, driven, at least in part, by advances in information technologies, that has led to a rise in U.S. labor productivity and to an increasingly integrated global economy. Since the 2001

---

[62]Aldrich, Howard and Martinez, Martha- "Entrepreneurship as Social Construction: An Evolutionary Approach." in Zoltan J. Acs and David B. Audretsch (Editors) - Handbook of Entrepreneurship Research: An Interdisciplinary Survey and Introduction, 2003

[63]Thornton, Patricia and Flynn, Katherine - Entrepreneurship, Networks, and Geographies, 2003

[64]Bathelt, Harald and Glucker, Johannes – The Relational Economy: Geographies of Knowing and Learning, 2011

recession, every one percent increase in productivity has eliminated 1.3 million jobs per year. Due to these structural changes in the U.S. economy, many of the jobs lost in the downturn are not coming back. Companies are able to do their present work with fewer people, as a result of the growth in information technology-based productivity. Automation and other productivity gains have been eliminating around seven million jobs per year. New jobs are not being created quickly enough because companies are postponing investing in new opportunities enabled by these same advances in technology, until they feel more confident about the future. This has led to our current, fairly unique condition: a jobless recovery. Job creation has to therefore become one of our top priorities.

We need to focus at least a portion of our innovation efforts to explore how these same great new technologies can be used to help us create new jobs, both in the short- and long-term. For example, recent estimates put today's world exports at $18.7 trillion, nearly twenty-five percent of world gross domestic product. According to the U.S. Department of Commerce less than one percent of the thirty million American businesses export. And of those, just over half export to more than one country. Export income supports the purchase of local goods and services, creating a "multiplier effect". Exporting, especially to international markets, entails high fixed costs and demands high firm productivity. As a result, exporting economies are overall more productive and wealthier. In the U.S., exporting firms pay 9.1 percent more than jobs in firms that export less and a dollar of exports produces twice as much employment as a dollar of domestic consumption[65]. Creation of public goods such as healthcare, education, infrastructure, and environmental protection is another way to provide more employment opportunities and sustain economic vibrancy[66].

Clusters promote knowledge-sharing and innovations in products and in technical and business processes by providing thick networks of formal and informal relationships across organizations. They are the incubators of innovation in their cohesive agglomeration of similar industries, technologies, skilled laborers, research institutions, and other necessary components. Rivals, customers, and suppliers all in one region promote efficiency and specialization, improvement and innovation. It increases concentration of knowledge, increases speed of information flow within national industry, and the rate at which innovations diffuse. The combination of national and intensely local conditions fosters competitive advantage which in advanced industries increasingly is determined by differential knowledge, skills, rates of innovation embodied in skilled people and

---

[65]Katz, Bruce and Bradley, Jennifer - The Metropolitan Revolution: How Cities and Metros are Fixing Our Broken Politics and Fragile Economy, 2013

[66]Gore, Al – The Future: Six Drivers of Global Change, 2013

organizational routines[67]. Thriving metropolitan economies such as Silicon Valley, Boston, New York, and Houston that carry multiple clusters essentially fuel the national economy through their pools of human capital, innovation, quality places for work and living, and infrastructure. Table 9.5 provides a more complete listing of such clusters.

Cities become drivers of innovation with their facilitative spaces and cradles of creativity. They become essential to the innovation system through the supply side: ready, available, abundant capital and labor; good infrastructure for productive activities; and diversified production structures that spawn synergies and hence innovation[68]. In addition they grow the innovation system on the demand side: diverse population of varying occupations, ideas, skills; high and differentiated level of consumer demand; and constant re-creation of urban order especially infrastructure of streets, water systems, energy, and transportation. Factors that bring clusters into being are difficult for government to control but the more important role for government is to enhance the competitiveness of existing and emerging clusters. In America, some examples of such emerging ecosystems are Austin's smart energy grids, Atlanta's re-shoring of manufacturing, Cleveland's retraining workers in community colleges, and North Carolina's high-tech exports.

Technology Innovation

Technology creates our world, our wealth, our economy, our very way of being. Innovation, an intimate of markets, that most mysterious of processes, is the accomplishment of economic tasks by novel means. Historians of technology, contemporary sociologists, and economic historians provide a fascinating tour of the literature on technological innovation, emphasizing the dreams and musings of inventors, novelists, and the popular media to show how they mediate new technological frames of references[69]. Following Schumpeter, scholars typically distinguish between invention, an idea made manifest, and innovation, ideas applied successfully in practice. In economics the change must increase value, either for the customer or the producer. Innovation and the introduction of it that leads to increased productivity is a fundamental source of increasing wealth in an economy. To become a world-class center of technological innovation, a society must have three basic elements – drive, a culture that supports change and hungers for it; human capital, the personal

[67]Porter, Michael - The Competitive Advantage of Nations, 1998

[68]Florida, Richard - The Flight of the Creative Class: The New Global Competition for Talent, 2007

[69]Flichy, Patrice - Understanding Technological Innovation: A Socio-Technical Approach, 2007

abilities that make world-class technology possible; and a capacity for mobilization, a society's ability to pursue ambitious new goals. ` Regimes and policies allowing for entrepreneurship and innovation, technological spillovers and externalities between collaborative firms, and systems that create innovative environments such as clusters, agglomerations, and metropolitan areas are also required[70].

We live in a society exquisitely dependent on science and technology, in which hardly anyone knows anything about science and technology[71]. Basic research is the pacemaker of technological progress[72]. If it is appropriately insulated from short-circuiting by premature considerations of use it will prove to be a remote but powerful dynamo of technological progress as applied R&D converts the discoveries of basic science into technological innovations to meet the full range of society's economic, defence, health and other needs. How multiple, unevenly paced and non-linear are the paths between scientific discovery and new technology; how often technology is the inspiration of science rather than the other way round; and how many improvements in technology do not wait upon science at all[73]. Universities, the intellectual hubs of the creative economy, are regional centers of innovation based on powerful networks linking its research and technology to industry, venture capital and entrepreneurs. The rate of return on basic science is about three times that of applied R&D, which, in turn, is about double the rate of return on physical capital[74]. Long-distance television transmission, photovoltaic solar cells, the transistor, the UNIX operating system, and cellular telephony - each of these innovations laid the groundwork for vibrant new industries. The transistor alone is the building block for computers, consumer electronics, telecommunications, high-tech medical devices, and much more. Likewise, the internet and the graphical user interface set the stage for the personal computer revolution, and unleashed cycles of applied innovation that created entirely new sectors of our economy.

Twenty years hence the seed corn that will create innovations then will have to be provided by the enormously expanded role of universities in the U.S. innovation system and not from big industry as was the case. Increasingly, universities will be the locus of fundamental discoveries. And industry will need to work with universities to transfer those discoveries into innovative products, commercialized through appropriate business models. Rapid business model prototyping is

---

[70]Wulf, William – The Innovation Ecology, 2008

[71]Sagan, Carl - Pale Blue Dot: A Vision of the Human Future in Space, 1997

[72]Bush, Vannevar – Science: The Endless Frontier, 1945

[73]Stokes, Donald – Pasteur's Quadrant: Basic Science and Technological Innovation, 1997

[74]Griliches, Zvi – R&D and Productivity: The Econometric Evidence, 1998

---

critically important to the future of technological innovation. The role of industry researchers should be expanded to include not just knowledge generation, but also knowledge brokering[75]. While only four percent of the nation's workforce is composed of scientists and engineers, this group disproportionately creates jobs for the economy. Failure to invest in the future now is likely to result in extended unemployment and a reduced standard of living for most Americans[76].

Only by providing leading-edge human and knowledge capital can America continue to maintain a high standard of living for its citizens. Eight studies conducted in recent decades indicate that public investments in science and technology have produced annualized societal returns that range from twenty to sixty-seven percent. Some economists estimate that about half the nation's growth in gross domestic product per capita during the last half-century can be attributed to scientific and engineering achievements. Given the increasing pace of advancement in science and technology, and their close companion, innovation, it seems likely that in the decades ahead these disciplines will have equal or greater impact[77]. One study in information technology concluded that only one new research "idea" in five hundred thousand results in a commercially profitable product. In addition to the implicit riskiness and uncertain applicability of investment in basic research, there is always the matter of its long-term nature, not uncommonly involving a decade or more of effort before results can be introduced into the marketplace. That constitutes a significant deterrent to investment by industry, which tends to have a "next quarter" focus.

Pursuits important to the public interest, and that have large societal returns as opposed to individual returns often of necessity become the province of government. The goal of government policy should be to make Americans more competitive, not to make American companies competitive. Government should not merely encourage supply of innovation, for example by funding research, but also try to stimulate demand. What we need is smart rather than pork-barrel industrial policy. The government should not seek to dictate, micromanaging every aspect of an agenda, but serve as a steward by convening and facilitating, spurring innovation by funding invention, pursuing smart educational policies including workforce retraining, creating and enforcing the right protections for intellectual property, focusing attention on flagging sectors of the economy such as manufacturing, enabling connections and alliances by supporting

[75]Chesbrough, Henry - Open Innovation: The New Imperative of Creating and Profiting from Technology, 2005

[76]National Academy of Sciences - Is America Falling off the Flat Earth?, 2007

[77]National Academy of Sciences - Rising Above the Gathering Storm: Energizing and Employing America for a Brighter Economic Future, 2007

networking and fostering key elements of infrastructure, for example broadband. It must practice strategic foresight, aggregate specialized expertise and knowledge, set up technical and business standards that promote collaboration, and serve as a hub for alliances. An entrepreneurial model for a government inured to bureaucracy is required. Individual states must have their own innovation strategy, after all they have to balance their budgets, they have to create local jobs, they have to educate people and children, they have to have progressive tax policies to attract companies, so it really is up to the governing executive branch in each state to act and not wait for handouts from the federal government or even directions[78].

Additionally, emerging economies are no longer content to be sources of cheap hands and low-cost brains. They too are becoming hotbeds of innovation, producing breakthroughs in everything from telecommunications to car-making to healthcare[79]. They are redesigning products to reduce costs not just by ten percent, but by up to ninety percent while the rich world is losing its leadership in the sort of breakthrough ideas that transform industries. Entrepreneurs in the developing world are applying the classic principles of division of labor and economies of scale to surprising areas such as heart surgery and cataract operations reducing cost without sacrificing quality. They are using new technologies such as mobile phones to bring sophisticated services to rural communities in everything from healthcare to banking. Innovation in the emerging world will encourage, rather than undermine, innovation in the rich world[80].

The more radical and revolutionary innovations tend to emerge from R&D, while more incremental innovations may emerge from practice. Products based on disruptive technologies are typically cheaper, simpler, smaller, and frequently more convenient to use, generally promise lower margins, not greater profits, are first commercialized in emerging or insignificant markets, and leading firms' most profitable customers generally do not want and indeed initially cannot use products based on disruptive technology[81]. The workforce is an important source of innovation. Another source of innovation, only now becoming widely recognized, is end-user innovation. This is where a person or company develops an innovation for their own personal or

---

[78]Fitzpatrick, Erika - Innovation America: A Final Report (National Governor's Association), 2007

[79]Friedman, Thomas – The World is Flat: A Brief History of the Twenty-first Century, 2005

[80]The Economist: A Special Report on Innovation in Emerging Markets (Masters of the New Management), 2010

[81]Christensen, Clayton - The Innovator's Dilemma: When New Technologies Cause Great Firms to Fail,1997

in-house use because existing products do not meet their needs[82]. Users play venturesome or entrepreneurial roles in the design of new products, bearing unmeasurable and unquantifiable risks and developing ground-level knowledge. The willingness and ability of users to undertake venturesome roles play a critical part in determining the ultimate value of innovations. The venturesome-ness of customers also encourages innovators to optimize their offerings for customers' needs and to invest in marketing to them. The traditional pattern of concentrating innovation-support resources on a few individuals is hugely inefficient. We must democratize the opportunity to create. Users, for example, were the developers of about eighty percent of the most important scientific instrument innovations, developers of most of the major innovations in semiconductor processing, and as were the developers of the most widely licensed chemical production processes[83].

The most powerful codes are now no longer in strings of ones and zeroes, but in four letters: A, T, C and G. These are the DNA-bases, the programming language of life. Different combinations of these four letters describe every life form on Earth. The ability to understand and manage them will revolutionize the competitive landscape in every sector, from medical to agricultural. The ability to treat disease by turning genes on and off may make some of today's medical therapies - amputations, consumption of toxic chemicals and irradiation - look as primitive as the practice of bloodletting to release evil spirits. With genomics, medicine can become predictive, preventive and personalized. And the industry will look considerably different when information technology capturing the DNA profile for every individual is as or more important as identifying chemical compounds with useful properties in treating disease. Exactly how biology, nanoscience and information science come together will drive much of future technology innovation.

New Job Creation

A competitive free-market system of innovation provides a positive but small share of gains to the innovator whereas users get the rest. Thus most of the benefits of technology change are passed on to consumers. Innovation can only sustain widespread prosperity if it benefits many users rather than a few innovators. The U.S. economy has emerged as the leading entrepreneurial economy in the world[84]. Engineering prowess combined with entrepreneurial spirit is our founding DNA, and central to American culture. Convincing evidence

---

[82]Hippel, Eric - Sources of Innovation, 1994

[83]Hippel, Eric - Democratizing Innovation, 2006

[84]Schramm, Carl - The Entrepreneurial Imperative: How America's Economic Miracle Will Reshape the World (and Change Your Life), 2006

---

exists that entrepreneurship is positively linked to economic performance[85].

The essence of entrepreneurship involves responsibility for uncertainty[86]. An important challenge faced by innovators is to persuade entrepreneurs to take a chance on their innovations in the absence of any hard evidence that returns are worth the risks. One notable feature of the U.S. innovation system is that a great many individuals and organizations are willing to be so persuaded. Entrepreneurship, neither science nor art, is the carrier of innovation. It is a practice. Innovation is the specific instrument used by entrepreneurs to exploit change, as an opportunity for a different business or a different service. An entrepreneur, somebody who upsets and disorganizes, offers an innovative solution to a frequently unrecognized problem[87]. Innovation has little to no value until it joins with entrepreneurship.

An inventor is a creative and purpose-based problem solver looking for better ways to do things. An inventor creates discoveries and breakthroughs, and is a rare thinker. The precious connector between an inventor and customers is an entrepreneur, the person who envisions a value and a customer and then creates a business model and strategy that create sales and profit. The business model really is everything. An entrepreneur is a person of action, optimism and determination. They get things done, they are rare doers. Entrepreneurship does not just provide supply but also builds demand, thus altering the supply-and-demand equilibrium. True entrepreneurs create new jobs and increase overall demand and spending. They take a current product or service and make it available to those who are not served or are underserved. They take a new idea and build enthusiasm, interest and desire for it, and thus create a new demand. For every thousand people, there are only three with the potential to develop an organization with fifty million dollars or more in annual revenues. There is an oversupply of discoveries, breakthroughs, and inventions, and an undersupply of entrepreneurship. We have overdeveloped the more controllable trait and left the more mysterious one's development to chance.

In terms of new job creation and total employee count, it is of interest to note from Table 2.1 that employment generated by vertically-integrated industrial companies far exceeds that of most newly-established internet-based, new economy companies – for example Xerox employs 130,000 employees when Facebook employs

[85]Keeble, David and Wilkinson, Frank – Collective Learning and Knowledge Development in the Evolution of Regional Clusters of High Technology SMEs in Europe, 1999

[86]Knight, Frank and McClure, John - Risk, Uncertainty, and Profit , 2009

[87]Drucker, Peter - Innovation and Entrepreneurship, 2006

approximately 10,000 while the two companies have comparable revenues.

Table 2.1   Employment Figures - Select U.S. Firms

Company	Industry	# of Employees	Year Started	Revenue $B (2014)	Revenue/ Employee
Walmart	Retail	2,200,000	1962	485.7	221,681
IBM	Computer Hardware & Software	379,592	1911	92.8	244,473
Amazon	e-commerce	224,400	1994	89.0	396,613
General Motors	Automotive	216,000	1908	155.9	721,759
Verizon	Telecom	177,300	1983	127.1	716,864
Zerox	Business Process Services	130,000	1906	19.5	150,000
Johnson & Johnson	Medical Equipment, Pharmaceuticals	126,500	1849	74.3	587,351
Microsoft	Computer Software & Hardware	118,584	1975	93.6	789,313
Intel	Semiconductors	106,700	1968	55.9	523,899
Apple	Consumer Electronics	115,000	1976	233.7	2,032,174
Pfizer	Pharmaceuticals	78,000	1849	49.6	635,897
ExxonMobil	Oil & Gas	75,300	1999 (Exxon 1882; Mobil 1911)	394.0	5,232,404
Google	Internet	59,976	1998	66.0	1,100,440
Motorola	Telecom	40,000	1928	11.9	297,500
Duke Energy	Electric Utility	29,250	1904	24.6	841,025
Next Era Energy	Renewable Energy	13,900	1925	7.0	503,597
Tesla Motors	Electric Transportation	12,000	2003	3.2	266,667
Facebook	Social Media	10,082	2004	12.5	123,983
Gilead Sciences	Biotechnology	7,000	1987	25.5	364,286
First Solar	Renewable Energy	5,600	1999	3.4	607,143
Uber	Transportation	?	2009	2.0	-
AirBnB	Travel	?	2008	0.9	-
TOTAL		**4,125,184**		**$2,028**	

Many players - entrepreneurs, managers, financiers, salespersons, consumers and not just a few brilliant scientists and engineers - have kept the U.S. at the forefront of the innovation game. The first version featured individual visionary inventors, the second the industrial model of innovation exemplified by the likes of Bell Labs, Xerox PARC, and HP Labs. IBM incubated or stimulated tens of thousands of great organizations including Microsoft and Intel; Google spins out hundreds of startups. Such highly inspired workplaces hatch thousands of new startups. There are now about twelve thousand corporate labs in the U.S. alone that spend about $150 billion annually, employing some seven hundred thousand scientists and engineers. Many of these big companies are scrapping large, centralized R&D teams in favor of smaller in-house units and alliances with promising start-ups. The innovation system then evolved to the innovator-entrepreneur model financed by venture capital[88]. Now we have innovative business models that may originate anywhere, driven by a global diffusion of innovation capability[89].

Material Science Challenges
Nanomechanical Characterization Technologies

This company based in Minneapolis, Minnesota has so far received 13 federal grants and contracts – five from the Department of Defense worth $1.6 million and eight from the Department of Energy worth $2.8 million. This company founded in1993 as a 3-person firm received its first federal funding in 1994, and has grown to a 125-person company today. It is the world leader in the development and commercialization of nanoscale test instruments and is a pioneer of in-situ imaging with nanomechanical property measurement capabilities. Its technical innovation was a patented one square centimeter 3-plate capacitive transducer that bridged the gap between existing qualitative atomic force microscopy and quantitative instruments available in the 1990s. As a result of this innovation, the Army was able to measure ultra-thin laser refractive coatings of canopies and face shields, the Navy was able to develop self-cleaning materials for ships' hulls, and NASA was able to repair the original problem with the coatings on the Hubble telescope. Other customers were able to test the 100 nanometer (nm) coatings on computer disk drives and the 10 nm coatings on Gillette Mach III razor blades. The company has sold its instruments in 60 countries and generated $225 million in revenue over the last 21 years of operation. The firm maintains that had it not been for federal funding, it would have simply sold its transducer technology into the commodity catalog sale market. The company's instrument capabilities include nanoindentation, tribology, modulus mapping, dynamic mechanical analysis, acoustic emission monitoring, electrical contact resistance, and in-situ scanning electron microscopy and transmission electron microscopy nanomechanical testing.

Startups are playing an increasingly important role in the U.S. economy; small businesses employ thirty percent of high-tech workers such as scientists, engineers, and information technology workers. Startups themselves, rather than silicon chips and disk drives, are becoming the raw materials of Silicon Valley. In recent decades, sixty to eighty percent of all newly created jobs have been created by small and medium enterprises (SME), although large companies remain critical because they are customers of SMEs, and do almost all the exporting[90]. For job creation, organic growth is better than acquired growth because mergers and acquisitions destroy jobs more than they create. The commercial success of a new product requires innovators to complement technical advantages with nontechnical development in

---

[88]Note: The five startups profiled in this chapter feature examples of job creation by publicly-funded high-tech small businesses.

[89]Kao, John - Innovation Nation: How America is Losing Its Innovation Edge, Why It Matters, and What We Can Do to Get It Back, 2007

[90]Clifton, Jim - The Coming Jobs War, 2011

areas such as organizational capabilities, marketing knowledge, and customer relationships. After business models, the U.S. needs to be best at innovation, entrepreneurship and global customer intelligence. For new job creation, consider that every one billion dollar invested in a new coal-fired plant yields eight hundred and seventy jobs; solar power

Shooting for the Stars
Multipurpose Variable-gravity Platform

This small firm based in Greenville, Indiana and in the business of space exploration has recently over a span of one year had two technologies fly in space, and a third acquired. The company began as a student project that was to fly on Challenger's STS-51L mission, and eventually flew on STS-29. The company has so far been awarded eighty-nine federally-funded grants and contracts – 43 from NASA worth $12 million, 33 from the Department of Defense worth $9.1 million, and 10 from NIH worth $9.3 million. When it started in 1989 it had two employees and now in 2015 has 40 employees – over the past five years, the firm has increased its personnel by 23 percent and its revenue by 33 percent. This small business applies a multidisciplinary approach to every project. Its technologies advance the understanding of microgravity and will help get astronauts safely to Mars, and will affect research for years to come onboard the International Space Station (ISS). Its Bone Densitometer is helping NASA understand microgravity's impact on human structure. The company's Life Science Research Sample Transfer Technology allows NASA astronauts to conduct on-orbit analysis in real time. This technology was actively used aboard the ISS when it flew on SpaceX-7 in June 2015. Their multi-purpose Variable-g Platform recently received a NASA Commercial Readiness Program award, which will lead to it being flown onboard the ISS in the future. The technology will be an enabling device for conducting microgravity research of high-value medical-grade materials.

plant one thousand nine hundred jobs; wind power plant three thousand three hundred jobs if turbines and blades are made here; energy efficiency through big-building retrofits seven thousand jobs; and home retrofits eight thousand jobs. Conservation therefore delivers the biggest bang for the buck - in less than a decade $520 billion in efficiency improvements will save $1.2 trillion and will cut consumption by twenty-three percent[91].

What is required are targeted investments for new job creation, a high-tech jobs program using technology to create high-wage jobs through inventions, entrepreneurs, and innovative business models for the new industries of the future such as space tourism, self-driving cars, personalized medicine, tissue engineering, precision agriculture, smart robots, alternative energy, and cancer cures. We need to transform startups into the sort of giants like Google and Cisco that create a greater proportion of high-paying jobs than those provided by startups. American startups can become global players in a generation.

---

[91]Clinton, Bill - Back to Work: Why We Need Smart Government for a Strong Economy, 2011

Despite the impressive U.S. record in pure innovation, it actually fails to translate into mass production and thus high employment industries as well as they do in Japan and China[92]. Fragmentation of U.S. high-

---

Collect Cellphones to Recycle
Kiosk Hardware for Automated Inspection

This startup, based in San Diego, California received two grants from the federal government worth $650,000 to design and commercialize a consumer self-serve, automated kiosk for the evaluation, buy back, and collection of used electronics directly from consumers. Prototype kiosks deployed during the initial R&D phase provided convincing proof of the feasibility of the baseline technical approach to the visual and electrical inspection technology, robotics, and the market. Financial metrics achieved were many multiples better than industry leading kiosks such as Coinstar or Redbox. Further R & D was required to achieve enough reliability in the automated inspection systems and the kiosk hardware to lead to the permanent removal of kiosk attendants in the field that currently serve as the fail-safe mechanism in current prototype systems. Broad commercial success relies on the development of a robust, designed-for-manufacturability (DFM), designed-for-serviceability (DFS), commercially reliable kiosk with a minimum retail field life of 5 years that incorporates needed improvements learned from the first phase R&D effort including refinements to the visual inspection system and algorithms, electrical inspection system, test station robotics subsystems, ergonomics, graphical user interface, and channel management systems. The company also hopes to further develop the system's capability to offer personal data erasure and expand accepted device types to potentially include digital cameras, portable game players, printer cartridges, laptops, eReaders, and tablets. The impact of this startup's patented system is that it finally achieved the threshold of consumer convenience and financial incentive required to inspire mass consumer participation in electronics recycling. Pilot market tests indicate that the company harvested 20 times more used phones than the next closest competitor in the test areas. As the company scales nationally it will divert mass amounts of toxic eWaste from our landfills, and place large sums of cash back in the hands of their customers and the retail locations hosting the kiosks, providing stimulus and incentive for these stakeholders to help forever alter the current wasteful lifecycle of consumer electronics. On average, each kiosk collects enough eWaste to offset its own annual energy consumption after just 5 days placement resulting in 360 days of CO_2 offset. An average each kiosk collects over 7,000 phones per year, which according to the EPA calculator is equivalent to taking the CO_2 emitted by 35 houses off the grid for a year. National and global media have already taken notice of this firm - the United Nation's Low Carbon Leadership Program recognized this small business as one of the best ideas in the world for the reduction of CO_2 on a global basis. The company started with 2 employees in 2008 and over the next five years had grown to 270 employees in 2013 when it was acquired for $350 million.

---

technology research into thousands of small companies may be optimal for innovation itself but it is not optimal for mass commercialization - indeed the side effect is that it is exceptionally easy for foreign companies to buy up U.S. innovations a la carte.

---

[92]Fletcher, Ian - Free Trade Doesn't Work: What Should Replace It and Why, 2011

Why does it matter that America not get pushed out of industries of the future? It is because having many increasing-returns industries is the only way to continue to be a developed economy. Such industries can absorb endlessly rising capital investment such as factory floor machinery, human capital or skills accumulation, and R&D. One big reason such industries are susceptible to innovation and R&D is their capacity for large capital absorption which activates a virtuous cycle – innovation absorbs capital, repays it by raising profitability which in turn generates more capital repeating the cycle. It is thus no accident that manufacturing and related fields generate over

---

Rockets into Space
Next-generation Missile and Launch Systems

This small business was founded in 2002 in Kirkland, Washington with ten employees and received its first small business innovation research (SBIR) award 8 months later. It has since then received 37 awards – 32 from the Department of Defense worth $11.1 million and 5 awards from NASA worth $1.3 million. Today, the company has 38 fulltime employees with federal funding continuing to play a crucial role by providing resources and inspiration for cutting edge R&D of advanced technologies for next-generation missile and space launch systems. The company's product line and credibility has increased, making it an established industry supplier of technologies such as stage separation systems, shroud/fairing deployment systems, space payload deployment, cryogenic pyro-valves, insensitive munitions technologies, and ionic salt-based "green" monopropellant fuel feed and ignition technologies. Company sales have grown from $1.6 million in 2004 to over $10 million in 2014. It is an employee-owned small business with innovative and motivated engineers and staff that provides its customers with a fast response from concept development to flight qualified products including end-to-end system testing. The firm's customers include the U.S. military, Boeing Defense Systems, Lockheed Martin, Raytheon Missile Systems, Northrop Grumman, Boeing, Orbital/ATK, and United Launch Alliance.

---

seventy percent of R&D in the U.S. and that within manufacturing high-tech accounts for roughly twenty percent of output but sixty percent of R&D. This susceptibility to innovation derives largely from the fact that "good" industries tend to produce goods capable of infinite improvement like laptops and airplanes while "bad" industries produce goods like t-shirts and fruits whose character is fixed. Products of "good" industries are also susceptible to a meaningful variety so firms do not end up selling the same exact product in pure head-to-head competition. This spares firms raw price competition that drives down profits, wages, and funds available for further investment. They instead compete on quality, reliability, reputation, marketing, service, product differentiation, special understanding of buyer needs, rapid innovation and managerial sophistication. This enables them to accumulate strongly entrenched competitive positions where vulnerability to pure price competition, crucially by foreign labor, is not a significant issue.

Predictions made for future areas of economic growth indicate that the U.S. worker must be willing to seek opportunities, acquire knowledge that is valuable across industries, and adapt to multiple jobs and the career shifts that necessarily accompany turbulent, innovation-based markets. How can policymakers maintain and speedup the continuing transition toward a more entrepreneurial economy[93]? Federal legislation accelerated the commercialization of innovations in universities through the Bayh-Dole Act of 1980, which granted universities exclusive control over inventions funded by the federal government and provided an important impetus to the creation of high-tech small business by earmarking a fixed percentage of federal R&D

Regenerative Biologic Therapies
Ocular Hygiene Solutions

This biotechnology company based in Doral, Florida is the industry leader in regenerative wound healing therapies. Founded in 1997, the firm serves the ophthalmology, optometry, musculoskeletal and wound care markets through its proprietary platform technologies. Its core products include amniotic membrane and umbilical cord-based tissue products which are processed utilizing the company's CryoTek process. Over 200,000 human implants have been conducted using this process. This small business received its first federal grant in 2003 when the company had fewer than 20 employees and $1.2 million in revenue. Today, the company has received 11 SBIR grants and has more than 180 employees with estimated annual revenues of $60 million in 2015. In 2013, the firm received $2.8 million from angel investors and $10 million from two venture capital funds in growth equity financing. The company's products AmnioGuard, PROKERA, and Cliradex all received SBIR grants during their development and now the combined yearly sales of these ophthalmic products total $15 million, representing 40% of the company's total revenue. To grow the company even further, the company has received recent SBIR awards that will support the company to move from human cells, tissues, and cellular and tissue-based products to drug and biologic products. It has so far received 12 awards from NIH worth a total of $6.1 million.

Collaborations with the University of Miami, the Walter Reed National Military Medical Center, the University of Manchester, the University of Texas, the University of Cincinnati, the University of Columbia, and many doctors' practices have allowed it to conduct research with new products or existing products in new indications. The company has pioneered stem cell research using human amniotic membrane to treat corneal surface conditions and developed Amniograft, the only tissue graft designated by the FDA as homologous for promoting ophthalmic wound healing while suppressing scarring and inflammation. The National Institute of Health has supported the company's research with more than 25 continuous years of research grants. The firm has just completed facility expansion of a state-of-the-art biotechnology manufacturing cleanroom facility.

---

funds for such businesses, under the Small Business Innovation Development Act of 1982. Also, America's enlightened treatment of bankrupt firms remains a model to the world - the best way to get more people to start businesses is to make it easier to wind them down.

---

[93]Kauffman Foundation - Roadmap for an Entrepreneurial Economy, 2006

The need to beat the Soviet Union was a decisive factor in disciplining the U.S. government to pursue an effective industrial policy. This can happen again with China. We have industrial policy for agriculture[94]! It is impossible not to have such a policy where the government helps only to plant long-term technology seeds in areas of private market failure or acute public need. Some of these technology seeds sprout, others not, but planting, the activity as a whole, must go forward if long-term economic gains are to be effectively harvested. The neglect of U.S. industrial policy has starved basic and applied research in critical areas such as bio-computing, optoelectronics, advanced materials, factory automation, energy conversion and storage, nano-manufacturing, and robotics.

America's increasingly patchy technology base renders it vulnerable to foreign suppliers of "key" or "chokepoint" technologies. For negative externalities like environmental damage we should tax imports and exports produced in environmentally harmful ways. Positive externalities like technology spillovers can be fixed by tax credits for R&D but without protection it can end up subsidizing research whose value is harvested by production abroad. In America political considerations quickly overwhelm economic merits, and industrial policy more closely resembles life support for dying industries than incubation of emerging ones. For example, semiconductor manufacturing is high-tech, and it is precisely this kind of factory that we should want to re-shore. They have scale economies that cause retention, high returns, high wages, and all the other effects of "good" industries.

---

[94]Friedman, Thomas - Hot, Flat, and Crowded: Why We Need a Green Revolution - and How It Can Renew America, 2009

Part II

Discovery to Innovation

Where is the knowledge we have lost in information?
- T.S. Eliot

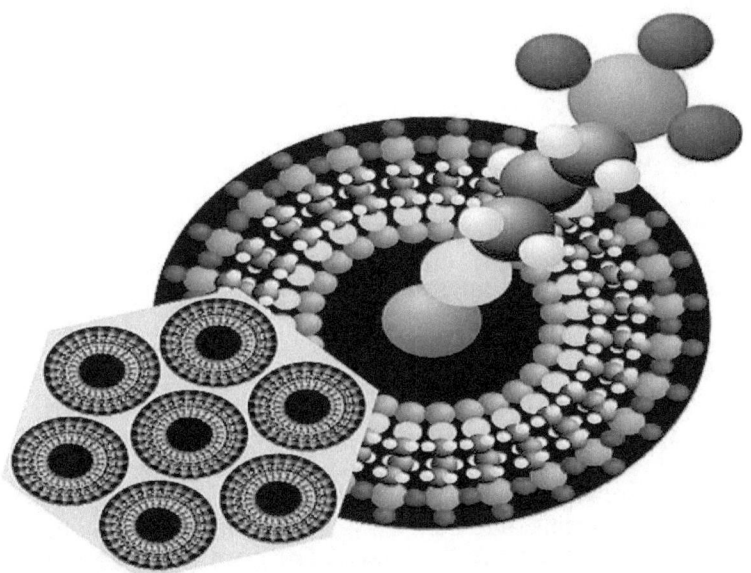

It is the duty of the sovereign or commonwealth to erect and maintain those public institutions and those public works, which, though they may be in the highest degree advantageous to a great society, are however of such a nature that the profit could never repay the expense to any individual or small number of individuals, and which it therefore cannot be expected that any individual or small number of individuals should erect or maintain[95]. The technological innovation that has driven much of the world's economic development since the Industrial Revolution would not have been possible without scientific knowledge.

Federal Government as Partner

Americans have long recognized the importance of science and technology for our prosperity, health, and security, and have invested substantial amounts of money in supporting the scientific enterprise, both privately and through government[96]. The conventional political wisdom regarding public financing for science is that basic scientific breakthroughs underpin commercial innovation but provide little or no direct profit themselves, so basic scientific research relies heavily on public support. Research funding is a term generally

---

[95]Smith, Adam - An Inquiry into the Nature and Causes of the Wealth of Nations, 1776

[96]Kennedy, Joseph -The Sources and Uses of U.S. Science Funding, 2012

covering any funding for scientific research obtained through a competitive process, in which potential research projects are evaluated and only the most promising receive funding. Such processes, which are run by government, corporations or foundations, allocate scarce funds.

The capitalist "market" has from the start been heavily shaped by government actions[97]. Most of the discoveries that have fueled modern industry in America were made in the fifties and sixties. Where then were the venture capitalists when all the funding had to be done in the basic sciences? The government actively paved the way for future industry development by bringing the courage, vision and funding so lacking in the private sector, by investing in new technology until fear-inducing uncertainty was transformed into mere risk. The government was seen as entrepreneur, risk-taker and market creator. Most radical new technologies in different sectors from pharmaceuticals to the Internet trace their funding to a courageous, risk-taking public sector. Venture capital has depended on government for the more expensive and uncertain research, before entering and cashing in when the uncertainty of investing in new innovations have been significantly reduced[98]. Key innovations would not have come about had we waited for the market and business to do it alone or have the government simply stand aside and provide the basics. The government, a key partner of the private sector, made bold decisions on the direction of change, and took on risks shaping and creating new markets.

In twentieth-century America there have been innumerable instances of public sector-seeded hundred billion dollar industries as indicated in Table 3.1. The pharmaceutical sector has been the biggest beneficiary of publicly-funded research. Most revolutionary new drugs are produced mainly with public, not private funds. The government undertakes the riskiest research, and it is the large pharmaceutical firms that then cash in major rewards. The development of the U.S. biotech industry is a direct product of the key role played by the National Institutes of Health and by extension the American taxpayer who has long been the nation's and the world's most important investor in medical knowledge creation. The Orphan Drug Act played a central and leading role in the development of biotechnology, just one of many critical moves the U.S. government made in supporting this industry. Wireless device technology provides another example. All key technologies behind the iPhone has been government-funded – communication technologies, Internet, the global positioning system, touch-screen display, the voice-activated personal assistant, and the

---

[97]Polanyi, Karl – The Great Transformation: The Political and Economic Origins of Our Time, 1944

[98]Mazzucato, Mariana - The Entrepreneurial State: Debunking Private v. Public Sector Myths, 2013

---

like. The government even had to support the commercialization of the Internet.

Table 3.1   Public Sector-seeded Industries

Advanced Batteries
Advanced Computer Graphics
Agriculture
Aircraft
Biotech, Genomics, Synthetic Biology
Cellular Phones, Smart Phones
Computers
Consumer Electronics
Databases, Expert Systems, Artificial Intelligence
Electronic Design Automation
Global Positioning System, Location-based Services
Internet
Medical Devices
Nanotechnology
Nuclear Energy
Optoelectronics
Genetics-based Drug Discovery
Robotics
Satellites
Semiconductors, Microprocessors

Around two-thirds of R&D in scientific and technical fields is carried out by industry, and twenty percent and ten percent respectively by universities and government. It is important for government not to do things which individuals are doing already, and to do them a little better or a little worse; but to do those things which at present are not done at all[99]. The U.S. government spends more than other countries on military R&D, although its proportion of such funding in the world has fallen from around thirty percent in the 1980s to under twenty percent. It spends approximately thirty-six percent of the world's total spending on medical research. With some exceptions like biotechnology, government provides the bulk of the funds for basic scientific research. In commercial R&D, all but the most research-oriented corporations focus more heavily on near-term commercialization possibilities rather than "blue-sky" ideas. Critics of basic research are concerned that research funding for the sake of knowledge itself does not contribute to a great return. However, scientific innovations often foreshadow or inspire further ideas unintentionally. An advantage to government-sponsored research is that the results are publicly shared, and can result in mass

---

[99]Keynes, John – The End of Laissez-Faire, 1926

collaborative projects that are beyond the scope of isolated private researchers whereas with privately-funded research the ideas are controlled by the private entity.

Table 3.2 provides the current sources of R&D funding in America[100]. Of all U.S. federal funds spent on science and technology R&D, about fifty-five percent goes to defense R&D and twenty percent to the National Institutes of Health. The National Science Foundation, the flagship of basic research for the U.S. government, gets five percent of these R&D funds. A brief description of the five U.S. federal agencies among others that fund basic research is provided in the next section.

Table 3.2   Sources of U.S. R&D Funding

R&D Funding Source	$Billions (2014)
Federal Government	123.0
Industry	307.5
Academia	13.3
Other Government	4.0
Non-profit	16.7
Total	**464.5**

An entrepreneurial public sector invests in areas that the private sector would not invest even if it had the resources. In biotechnology, nanotechnology, and the Internet, private venture capital arrived fifteen to twenty years after the most important investments were made by the public sector. Government funding for early-stage technology firms is equal to the total investments of business angels and about two to eight times the amount invested by private venture capital[101]. Shareholders and executives are rewarded for riding the innovation wave that the government created. The history of technological change teaches us that choosing particular sectors is absolutely crucial. The government's role is not just to create knowledge through national labs and universities, but also to mobilize resources that allow knowledge and innovation to diffuse broadly across sectors of the economy.

It is imperative to therefore consider both laissez faire free-market principles *and* industrial policy. For example, going way beyond simply funding research, the Defense Advanced Research Projects Agency funded the formation of computer science departments, provided startup firms with early research support, contributed to

---

[100]www.battelle.org – 2014 Global R&D Funding Forecast, 2013

[101]Auerswald, Philip and Branscomb, Lewis - Valleys of Death and Darwinian Seas, 2003

semiconductor research, supported human-computer interface research, and oversaw the early stages of the Internet. Agency officers engaged in business and technological brokering by linking university researchers to entrepreneurs interested in starting a new firm, connecting startup firms with venture capital, and finding a larger company to commercialize the technology or assisting in procuring a government contract to support the commercialization process. In the case of nanotechnology, the next general purpose technology, the U.S. government has been the lead visionary in dreaming up the possibility of a nanotech revolution, another example of going beyond simply creating the right infrastructure, funding basic research, and setting rules and regulations.

Decades of government investment in the science and technology base have made America a successful innovator, but the country has failed to secure high levels of employment, to increase tax revenues, and to promote export of goods and services. Corporations are increasingly reluctant to provide public goods that spilled over to society from their industrial labs because they cannot capture the full rent from R&D. The wedge between private and social returns arising from R&D spillovers is just as true in the era of Bell Labs as they are now. What is missing most today is the private component of R&D working in real partnership with the public component creating a more symbiotic ecosystem. As firms like Apple, Google, GE, and Cisco flourish financially the U.S. economy struggles to find its way out of debilitating economic issues like the growing trade deficit against Asian economies, declining manufacturing activities, increasing unemployment, widening budget deficits, inequality, and a deteriorating infrastructure.

Venture capitalists have convinced policymakers and much of mainstream media that they are the "entrepreneurial" force in the knowledge economy. We need to better identify how the division of "innovative labor" maps into division of rewards. Reaping proper returns is crucial because innovation cycles can then be sustained over time. Most thinkers on strategic management and organizational change have focused more on the private sector, leaving the public sector to simply focus on "creating the conditions" for innovation to happen in the "revolutionary" private sector. We need to transform the public sector from within so that it is more strategic, meritocratic and dynamic. While government spending on basic education and health should not necessarily expect a direct return beyond the taxes and supply of skilled and healthy staff, the government's high-risk investments should be thought of differently, and allowed to reap a direct return precisely because the failure rate is so high. While privatization of gains and socialization of losses in the financial sector are recognized as economically inefficient and socially unjust, the same asymmetry for new-tech firms and for more mature firms that need external investment to turnaround, has remained unnoticed.

Federal Basic Research Funding

No single factor is more important to the intellectual and economic progress of society, and to the enhanced well-being of its citizens, than the continuous acquisition of new knowledge. It is generally been accepted that science can be divided into three main categories. Basic research is research that seeks to gain more complete knowledge or understanding of the fundamental aspects of phenomena and of observable facts, without having in mind applications toward specific processes or products. How did biology begin? How did living things get going and do they elsewhere than Earth? What caused the Cambrian explosion, a mere five hundred and forty-two million years ago when animal life suddenly took off? The most profound scientific mystery of all: Consciousness, one that defines what it means to be human. Why does time pass? The equations of physics suggest time should be able to go backwards as well as forwards. Experience suggests, though, that it cannot. Why? And is time travel really possible? Is the universe alone? Would the world make more sense if other universes existed? What is the universe really made of? What is it that makes up ninety-five percent of the cosmos?

Applied research, on the other hand, is aimed at knowledge necessary to determine the means by which a recognized need may be met. Development, in turn, is defined as the systematic use of the knowledge or understanding gained from research, directed toward the production of useful materials, devices, systems, or methods, including design and development of prototypes and processes. These three forms of work in science and technology are collectively referred to as research and development. They represent the different stages of the innovation pathway, from scientific breakthroughs to the creation of useful products.

National Institutes of Health

The mission of the National Institutes of Health (NIH)[102] is to seek fundamental knowledge about the nature and behavior of living systems and the application of that knowledge to enhance health, lengthen life, and reduce illness and disability. The goals of the agency are to foster fundamental creative discoveries, innovative research strategies, and their applications as a basis for ultimately protecting and improving health; to develop, maintain, and renew scientific human and physical resources that will ensure the nation's capability to prevent disease; to expand the knowledge base in medical and associated sciences in order to enhance the nation's economic well-being and ensure a continued high return on the public investment in

---

[102]www.nih.gov

research; and to exemplify and promote the highest level of scientific integrity, public accountability, and social responsibility in the conduct of science.

This federal agency is made up of twenty-one Institutes and six Centers, as shown in Table 3.3, each with a specific research agenda, often focusing on particular diseases or body systems. In realizing

Table 3.3   NIH Institutes and Centers

National Cancer Institute
National Eye Institute
National Heart, Lung, and Blood Institute
National Human Genome Research Institute
National Institute on Aging
National Institute on Alcohol Abuse and Alcoholism
National Institute of Allergy and Infectious Diseases
National Institute of Arthritis and Musculoskeletal and Skin Diseases
National Institute of Biomedical Imaging and Bioengineering
National Institute of Child Health and Human Development
National Institute on Deafness and Other Communication Disorders
National Institute of Dental and Craniofacial Research
National Institute of Diabetes and Digestive and Kidney Diseases
National Institute on Drug Abuse
National Institute of Environmental Health Sciences
National Institute of General Medical Sciences
National Institute of Mental Health
National Institute on Minority Health and Health Disparities
National Institute of Neurological Disorders and Stroke
National Institute of Nursing Research
National Library of Medicine
Center for Information Technology
Center for Scientific Review
Fogarty International Center
National Center for Complementary and Integrative Health
National Center for Advancing Translational Sciences
NIH Clinical Center

these goals, the NIH provides leadership and direction to programs designed to improve the health of the nation by conducting and supporting research in the causes, diagnosis, prevention, and cure of human diseases; in the processes of human growth and development; in the biological effects of environmental contaminants; in the understanding of mental, addictive and physical disorders; and in directing programs for the collection, dissemination, and exchange of information in medicine and health, including the development and support of medical libraries and the training of medical librarians and other health information specialists. The agency invests nearly $30 billion annually in medical research for the American people. More than eighty percent of its funding is awarded through almost 50,000 competitive grants to more than 300,000 researchers at more than

2,500 universities, medical schools, and other research institutions in every state and around the world. About ten percent of its budget supports projects conducted by nearly 6,000 scientists in its own laboratories, most of which are on the NIH campus in Bethesda, Maryland.

An example of basic research funding by this agency is to the Intellectual and Developmental Disabilities Research Center (IDDRC) at the University of Pennsylvania (Penn) and the Children's Hospital of Philadelphia (CHOP). This Center, now in its 25th year, is an interdisciplinary program that is the chief entity at CHOP/Penn for the promulgation of research into the causes and treatment of developmental disabilities.

With this funding, the Center will build upon its foundation through the following initiatives: A research project which uses magnetoencephalography to develop a novel biomarker - an "electronic signature" - for non-verbal and minimally verbal children with autistic spectrum disorders; a series of six research core facilities which will provide investigators at CHOP/Penn with state-of-the-art facilities and expertise, including two new cores, preclinical models and clinical translational; an educational program which will nurture and create an "IDD Community" by featuring monthly seminars, including the IDDRC Research Lecture, the monthly "Chalk Talks" and the Autism Distinguished Lecture Series; a career development initiative which will benefit dozens of exceptional young investigators who will receive support from the Center's new program development award (to be funded by the Philadelphia Foundation), the Center-administered T32 training grant in neurodevelopmental disabilities and several private awards (aggregate value approximately $3 million over 5 years) given by local philanthropies to the Center; a research advocacy mission which involves an IDDRC- CHOP/Penn collaboration to create "centers of excellence" that encourage inter-disciplinary translational research into IDD; and participate in the larger Center network in order to realize the scientific and organizational goals of the IDD Branch of the National Institute of Child Health and Human Development (NICHD).

The Center will support seventy-eight federally-funded projects ($23.3 million/year), of which fifteen projects are from NICHD ($5.8 million/year). The theme of this IDDRC is "Genes, Brain, Behavior", a designation which reflects ongoing efforts to understand IDD in three inter-related domains: The genetic anlage which causes a disability or modulates disease severity; the anomalies of brain biochemistry and neurophysiology which accompany genetic changes; and the phenomenological manifestations of these genetic/neurophysiologic alterations which are recognized as clinical manifestations of IDD. The rationale for this "reductionist" approach is that the dissection of a disability into isolated and "simpler" components affords a strategy with which to develop therapies that will prevent, attenuate or even reverse the devastating consequences of the disorder.

Department of Energy

The Department of Energy (DOE)[103] is entrusted with critical responsibilities for America's security and prosperity by sustaining the nuclear deterrent without testing, supporting the Navy nuclear propulsion program, and in an age of global terrorism, committed to controlling and eliminating nuclear materials worldwide. Its responsibilities include enabling the transition to a low-carbon secure energy future, especially by developing low-cost energy technologies and a 20th century resilient energy infrastructure. It supports America's research community, especially in the physical sciences, as the foundation for discovery and innovation. It also protects public health and safety through a long-term commitment to cleaning up waste from nuclear weapons production, and by developing and maintaining emergency response capabilities for nuclear and radiological incidents and energy infrastructure disruptions.

Founded during the immense investment in scientific research in the period preceding World War II, the National Laboratories have served as the leading institutions for scientific innovation in the United States for more than sixty years. The seventeen DOE national laboratories, displayed in Table 3.4, are the core scientific and technical assets for carrying out these missions. They tackle the critical scientific challenges of our time, from combating climate change to discovering the origins of our universe, and possess unique instruments and facilities, many of which are found nowhere else in the world. They address large-scale, complex research and development challenges with a multidisciplinary approach that places an emphasis on translating basic science to innovation. Specifically, these laboratories conduct research of the highest caliber in physical, chemical, biological, and computational and information sciences that advances our understanding of the world around us. They advance U.S. energy independence and leadership in clean energy technologies to ensure the ready availability of clean, reliable, and affordable energy.

The National Labs enhance global, national, and homeland security by ensuring the safety and reliability of the U.S. nuclear deterrent, helping to prevent the proliferation of weapons of mass destruction, and securing the nation's borders. They also design, build, and operate distinctive scientific instrumentation and facilities, and make these resources available to the research community. The National Labs house five of the world's ten fastest supercomputers.

---

[103]www.energy.gov

Science is not linear, nor is it uniform, but the Lab system makes the pursuit of discovery, and the many solutions that result from it, both a collaborative enterprise and a shared national resource. The budget for this federal agency for fiscal year 2015 is $27.4 billion.

Here is an example of basic research funding selected by the DOE Office of High Energy Physics: This basic research funding was provided to a professor in the Department of Physics at Duke University to develop a tool to search for new physics through the study of coherent neutrino-nucleus scattering. At first blush, the fundamental particle that we call the neutrino does not interact with anything, does not weigh anything, and cannot be observed. Predicted in 1930 to explain the apparent violation in nuclear beta decay of the basic physics precept of energy conservation, the neutrino eluded detection for over two decades. The "little neutral one," as Enrico Fermi named it, can be difficult to work with but has traditionally offered vistas

Table 3.4   DOE National Labs

Ames Laboratory
Argonne National Laboratory
Brookhaven National Laboratory
Fermi National Accelerator Laboratory
Idaho National Laboratory
Lawrence Berkeley National Laboratory
Lawrence Livermore National Laboratory
Los Alamos National Laboratory
National Energy Technology Laboratory
National Renewable Energy Laboratory
Oak Ridge National Laboratory
Pacific Northwest National Laboratory
Princeton Plasma Physics Laboratory
Sandia National Laboratory
Savannah River National Laboratory
SLAC National Accelerator Laboratory
Thomas Jefferson National Accelerator Facility

to a better understanding of the physical world. The coherent scattering of neutrinos off nuclei is one such vista. Predicted over 40 years ago, it has long been one of the "hard" problems in neutrino physics. The daunting technological challenge and the dearth of suitable sources of neutrinos have so far stymied efforts to observe this process. The situation is rapidly changing with new developments in detector technologies and with the construction of the Spallation Neutron Source (SNS) at Oak Ridge National Laboratory. Research collaboration has recently endeavored to measure this process for the first time at the SNS. A precision measurement of coherent neutrino scattering can test for extensions to current models of the weak force; it is a sensitive test for large neutrino magnetic moments which can arise in theories with large extra dimensions; it can investigate the dependence of the interaction of the neutrino on different types of quarks; and it is perhaps the most natural way to search for sterile neutrinos, a flavor of neutrino which may be even more weakly interacting than its cousins.

An elegant and simple detector concept will be developed that incorporates numerous nuclear targets as swappable, low-threshold, kilogram-scale gas ionization detectors. The detectors provide for a robust, systematically clean and information rich suite of experiments. Coherent neutrino-nucleus scattering is a long-awaited tool that will enable sensitive searches for fundamental particle physics beyond our current understanding.

National Aeronautics and Space Administration

The vision of the National Aeronautics and Space Administration (NASA)[104] is to reach for new heights and reveal the unknown for the benefit of humankind. To do that, thousands of people have been working around the world, and off of it, for more than fifty years, trying to answer some basic questions. What is out there in space? How do we get there? What will we find? What can we learn there, or learn just by trying to get there, that will make life better here on Earth? This federal agency was established in 1958, partially in response to the Soviet Union's launch of the first artificial satellite the previous year. It grew out of the National Advisory Committee on Aeronautics (NACA), which had been researching flight technology for more than forty years. President Kennedy focused NASA and the nation on sending astronauts to the moon by the end of the 1960s. Through the Mercury and Gemini projects, this agency developed the technology and skills it needed for the journey. On July 20, 1969, Neil Armstrong and Buzz Aldrin became the first of twelve men to walk on the moon. Meanwhile, it was continuing the aeronautics research pioneered by NACA. Its aeronautics teams are working with other government organizations, universities, and industry to fundamentally improve air transportation and retain our nation's leadership in global aviation.

It also conducted purely scientific research and worked on developing applications for space technology, combining both pursuits in developing the first weather and communications satellites. After Apollo, the agency focused on creating a reusable ship to provide regular access to space: the space shuttle. First launched in 1981, this vehicle flew more than hundred and thirty successful missions before being retired in 2011. In 2000, the United States and Russia established a permanent human presence in space aboard the International Space Station (ISS), a multinational project representing

---

[104]www.nasa.gov

65

the work of fifteen nations. NASA also has continued its scientific research. In 1997, Mars Pathfinder became the first in a fleet of spacecraft that have been exploring Mars, as we try to determine whether life ever existed there. The Terra, Aqua and Aura Earth Observing System satellites are flagships of a different fleet, this one in Earth orbit, designed to help us understand how our home world is changing. Throughout its history, this agency has conducted or funded research that has led to numerous improvements to life here on Earth. In the early 21st century, NASA is extending our senses to see the farthest reaches of the universe, while pushing the boundaries of human spaceflight farther from Earth than ever before.

On Earth and in space, the agency is developing new capabilities to send future human missions to an asteroid and Mars. Mars once had conditions suitable for life. Future exploration on our Journey to Mars could uncover evidence of past life, answering one of the fundamental mysteries of the cosmos: Does life exist beyond Earth? The Journey to Mars begins aboard the ISS, where astronauts are extending permanent human presence in space and performing research that will help us understand how humans can live and work off Earth for long periods. To send astronauts deeper into the solar system, NASA is developing the most advanced rocket and spacecraft ever designed. Its Orion spacecraft will carry four astronauts to missions beyond the moon, launched from Florida aboard the Space Launch System, an advanced heavy-lift rocket that will provide an entirely new national capability for human exploration beyond Earth's orbit. To help test other spaceflight capabilities to meet the goal of sending humans to Mars, including advanced propulsion and spacesuits, NASA is developing the Asteroid Redirect Mission, the first-ever mission to identify, capture and redirect a near-Earth asteroid to a stable orbit around the moon, where astronauts will explore it in the 2020s, returning with samples. Ten field centers and a variety of installations, as shown in Table 3.5, conduct the agency's day-to-day work, in laboratories, on air fields, in wind tunnels and in control rooms.

An unprecedented array of science missions is seeking new knowledge and understanding of Earth, the solar system and the universe. We are studying Earth right now through current and future spacecraft helping answer critical challenges facing our planet: climate change, sea level rise, freshwater resources and extreme weather events. Multiple NASA missions are studying our sun and the solar system, unraveling mysteries about their origin and evolution. By understanding variations of the sun in real-time, we can better characterize space weather, which can impact exploration and technology on Earth. The New Horizons spacecraft flew by Pluto in July 2015, providing the closest views we have ever had of the dwarf planet. The Juno spacecraft, poised to reach Jupiter in 2016, will peer beneath its dense gas to reveal the mysteries of its core. NASA telescopes also are peering into the farthest reaches of the universe and back to its earliest moments of existence, helping us understand the universe's origin, evolution, and destiny. For more than twenty-five years, the Hubble Space Telescope (HST) continues to explore as the agency develops its successor, the James Webb Space Telescope, which will capture light from the universe's earliest stars. NASA's budget for fiscal year 2015 is $17.6 billion.

Table 3.5   NASA Centers and Facilities

Ames Research Center
Armstrong Flight Research Center
Glenn Research Center
Goddard Space Flight Center
Goddard Institute of Space Studies
Independent Verification and Validation Facility
Jet Propulsion Laboratory
Johnson Space Center
Kennedy Space Center
Langley Research Center
Marshall Space Flight Center
Michoud Assembly Facility
NASA Engineering and Safety Center
NASA Safety Center
NASA Shared Services Center
Plum Brook Station
Stennis Space Center
Wallops Flight Facility
White Sands Test Facility

An example of basic research funding by NASA is the use of the HST to gain a better understanding of the cosmos: Astronomers using this telescope have assembled a comprehensive picture of the evolving universe, among the most colorful deep space images ever captured by the 24-year-old telescope. Researchers say the image, in a new study called the ultraviolet coverage of the Hubble Ultra Deep Field (HUDF), provides the missing link in star formation. This image is a composite of separate exposures taken from 2003 to 2012 with Hubble's Advanced Camera for Surveys and Wide Field Camera 3. Astronomers previously studied the HUDF in visible and near-infrared light in a series of images captured from 2003 to 2009. The HUDF shows a small section of space in the southern-hemisphere constellation Fornax.

Now, using ultraviolet light, astronomers have combined the full range of colors available to Hubble, stretching all the way from ultraviolet to near-infrared light. The resulting image made from eight hundred and forty-one orbits of telescope viewing time contains approximately ten thousand galaxies, extending back in time to within a few hundred million years of the Big Bang. Prior to this study of the universe, astronomers were in a curious position. Missions such as

NASA's Galaxy Evolution Explorer observatory, which operated from 2003 to 2013, provided significant knowledge of star formation in nearby galaxies. Using Hubble's near-infrared capability, researchers also studied star birth in the most distant galaxies, which appear to us in their most primitive stages due to the significant amount of time required for the light of distant stars to travel into a visible range. But for the period in between, when most of the stars in the universe were born, a distance extending from about 5 to 10 billion light-years, they did not have enough data.

"The lack of information from ultraviolet light made studying galaxies in the HUDF like trying to understand the history of families without knowing about the grade-school children. The addition of the ultraviolet fills in this missing range," said the principal investigator at the California Institute of Technology. Ultraviolet light comes from the hottest, largest and youngest stars. By observing at these wavelengths, researchers get a direct look at which galaxies are forming stars and where the stars are forming within those galaxies. Studying the ultraviolet images of galaxies in this intermediate time period enables astronomers to understand how galaxies grew in size by forming small collections of very hot stars. Because Earth's atmosphere filters most ultraviolet light, this work can only be accomplished with a space-based telescope. "Ultraviolet surveys like this one using the unique capability of Hubble are incredibly important in planning for NASA's James Webb Space Telescope," said a team member at Arizona State University in Tempe. "Hubble provides an invaluable ultraviolet light dataset that researchers will need to combine with infrared data from Webb. This is the first really deep ultraviolet image to show the power of that combination."

The HST is an international project between NASA and the European Space Agency. NASA's Goddard Space Flight Center in Greenbelt, Maryland, manages the telescope. The Space Telescope Science Institute in Baltimore, operated for NASA by the Association of Universities for Research in Astronomy based in Washington D.C., conducts Hubble science operations.

National Science Foundation

The National Science Foundation (NSF)[105] is an independent federal agency created by Congress in 1950 "to promote the progress of science; to advance the national health, prosperity, and welfare; to secure the national defense". With an annual budget of $7.4 billion in fiscal year 2015, it is the funding source for approximately twenty-one percent of all federally-supported basic research conducted by America's colleges and universities. In many fields such as

---

[105] www.nsf.gov

mathematics, computer science, and the social sciences, NSF is the major source of federal backing. This federal agency fulfills its mission chiefly by issuing limited-term grants, currently about 11,000 new awards per year, with an average duration of three years, to fund specific research proposals that have been judged the most promising. Most of these awards go to individuals or small groups of investigators. Other awards provide funding for research centers, instruments, and facilities that allow scientists, engineers, and students to work at the outermost frontiers of knowledge. NSF's goals of discovery, learning, research infrastructure, and stewardship provide an integrated strategy to advance the frontiers of knowledge, cultivate a world-class, broadly inclusive science and engineering workforce, expand the scientific literacy of all citizens, build the nation's research capability through investments in advanced instrumentation and facilities, and support excellence in science and engineering research and education.

NSF is "where discoveries begin" - many of the scientific and technological advances that this agency has supported have been truly revolutionary. As of 2013, NSF-funded researchers have won two hundred and twelve Nobel Prizes. These pioneers have included the scientists or teams that discovered many of the fundamental particles of matter, analyzed the cosmic microwaves left over from the earliest epoch of the universe, developed carbon-14 dating of ancient artifacts, decoded the genetics of viruses, and created an entirely new state of matter called a Bose-Einstein condensate. This federal agency also funds equipment needed by scientists and engineers that is often too expensive for any one group or researcher to afford. Major research equipment include its national observatories with their giant optical and radio telescopes, Antarctic research sites, high-end computer facilities and ultra-high-speed network connections, ships and submersibles for ocean research, sensitive detectors of subtle physical phenomena, and gravitational wave observatories. Table 3.6 is a listing of engineering research centers, an example of the many kinds of basic research centers funded by NSF.

Another essential element of NSF's mission is support for science and engineering education, from pre-K through graduate school and beyond. The research funds are thoroughly integrated with education to help ensure that there will be enough skilled people available to work in new and emerging scientific, engineering and technological fields, and capable teachers to educate the next generation. This agency is also responsible for assessing the state of the scientific enterprise in America. To this end, it conducts surveys and studies published as reports, the most comprehensive of which is the biennial report which gives a broad base of quantitative information on U.S. and international science and engineering indicators[106]. These reports are a rich trove of information and statistics not only about the sources of funding for scientific research in America, but also about

---

[106]National Science Board Science and Engineering Indicators, 2015

American science education, the public's attitudes toward science, the science and engineering labor force, and other important indicators. For the purposes of tracking spending on scientific work, NSF divides science into three main categories - basic research, applied research, and development.

Table 3.6   NSF-funded Engineering Research Centers

Center for Structured Organic Particulate Systems
Center for Compact and Efficient Fluid Power
Center for Biorenewable Chemicals
Nanomanufacturing Systems for Mobile Computing and Mobile Energy Technologies
Synthetic Biology Engineering Research Center
Advanced Self-Powered Systems of Integrated Sensors and Technologies
Center for Sensorimoter Neural Engineering
Quality of Life Technology
Revolutionizing Metallic Biomaterials
Ultra-wide Area Resilient Electric Energy Transmission Networks
Future Renewable Electric Energy Delivery and Management Systems
Quantum Energy and Sustainable Solar Technology
Reinventing Nation's Urban Water Infrastructure
Smart Lighting ERC
Center for Integrated Access Networks
Mid-Infrared Technologies for Health and the Environment
Translational Applications of Nanoscale Multiferroic Systems

Historically, NSF's survey data for corporate R&D spending did not include companies with fewer than five employees. The agency has recently begun to address this omission with a microbusiness innovation science and technology survey. The smallest companies for which data are available, those with between five and twenty-four employees, spent an average of fourteen percent of sales on R&D, while the largest, those with over twenty-five thousand employees, spent 2.3%. As one would expect, industry tends to invest in the development stage of the innovation pathway. However, while the federal government is the primary supporter of basic scientific research, it too provides more money for the development stage than for research. Here are three examples of basic research funding by NSF:

Many of Alaska's more than hundred and thirty volcanoes are located along the 1,550-mile long Aleutian Arc. It extends from the Alaska mainland west toward Kamchatka, Russia, and forms the northern part of the tectonically active "ring of fire" girding the Pacific Ocean basin. To learn more about volcanic hazards in this volatile region, twenty-five researchers from eleven institutions studied the Aleutian Arc the summer of 2015. The effort is part of the NSF-funded Geodynamic Processes at Rifting and Subducting Margins (GeoPRISMS) program designed to investigate the architecture, mechanics and plumbing of continental margins at subduction zones and continental rifts, including what controls geohazards such as earthquakes and volcanoes. This project covers an unprecedented number of volcanic islands in this chain. The research will address volcanic systems, from magma storage to the chemistry and style of eruptions, as well as the earthquake and tsunami hazards that can affect the entire Pacific. Earlier in 2015, a magnitude 6.9 earthquake rocked parts of the Aleutians; there was no tsunami threat. Of particular concern: more than thirty thousand people each day, and eighty thousand large planes each year, fly through the skies over Aleutian volcanoes - mostly on heavily traveled great-circle routes to and from North America, Europe and Asia.

Bioluminescent organisms produce light through chemical reactions involving hundreds of various unrelated enzymes and other light-producing proteins that evolved independently in various groups of organisms. Researchers understand the biochemical reaction that produces light in only a handful of species. We know the most about how light is produced by fireflies. In addition, new bioluminescent creatures are still frequently being discovered, particularly on land and sea in tropical areas, and most often in coral reefs, which are among the most diverse ecosystems on Earth. With so much still to discover about bioluminescence, various aspects of this phenomena form the bases of vibrant and active fields that are currently being researched by biochemists, entomologists, marine biologists and engineers. Research into bioluminescence has already produced many practical applications such as helping to revolutionize our understanding of cancer by literally shining a light on what researchers could not see any other way. Scientists can make particular types of cells in laboratory animals bioluminescent, including cancer cells. Using high-tech imaging tools, researchers can track the movement of these bioluminescent cells by following the light they emit. This technique enables researchers to view proliferating and metastasizing bioluminescent cancer cells in real time, and thereby gain important insights about how cancer grows. It is revolutionizing how we study cells. An NSF-funded biologist's search for the source of the green glow of the jellyfish *Aequorea victoria* led him to discover a protein called green fluorescent protein which is now widely used in biological and biomedical research as a fluorescent tag.

A relentless global effort to shrink transistors has made computers continually faster, cheaper and smaller over the last forty years. This effort has enabled chipmakers to double the number of transistors on a chip roughly every eighteen months, a trend referred to as Moore's Law. In the process, the U.S. semiconductor industry has become one of the nation's largest export industries, valued at more than $65 billion a year. Transistor size will continue to decrease for a decade, reaching approximately five nanometers long and one nanometer (or about 5 atoms) wide in its critical active region. At this nanoscale, new phenomena take precedence over those that hold sway in the macro-world. Quantum effects such as tunneling and atomistic disorder dominate the characteristics of these nanoscale devices. Fundamental questions about how various materials and configurations behave at this scale need to be answered. Researchers at Purdue University developed primary software tools used by academics, semiconductor companies and students to predict the future behavior of nanoscale transistors. This software simulates the multiscale, multiphysics phenomena that occur when an electric charge passes through a few-atoms-wide transistor. In doing so, it helps researchers design future generations of nanoelectronic devices, including transistors and quantum dots, even before they can be physically produced, and predicts device performances and phenomena that researchers otherwise could not explore.

Defense Advanced Research Projects Agency

For more than fifty years, the Defense Advanced Research Projects Agency (DARPA)[107] has held to a singular and enduring mission: to make pivotal investments in breakthrough technologies for national security. The genesis of DARPA dates to the launch of Sputnik in 1957, and a commitment by the United States that, from that time forward, it would be the initiator and not the victim of strategic technological surprises. Working with innovators inside and outside of government, DARPA has repeatedly delivered on that mission, transforming revolutionary concepts and even seeming impossibilities into practical capabilities. The ultimate results have included not only game-changing military capabilities such as precision weapons and stealth technology, but also such icons of modern civilian society such as the Internet, automated voice recognition and language translation, and Global Positioning System receivers small enough to embed in myriad consumer devices. DARPA explicitly reaches for transformational change instead of incremental advances. But it does not perform its engineering alchemy in isolation. It works within an innovation ecosystem that includes academic, corporate and governmental partners, with a constant focus on the nation's military services, which work with DARPA to create new strategic opportunities and novel tactical options. For decades, this vibrant, interlocking ecosystem of diverse collaborators with about two hundred and fifty ongoing R&D programs has proven to be a nurturing environment for the intense creativity that DARPA is designed to cultivate.

DARPA has owned the critical mission of keeping the United States out front when it comes to cultivating breakthrough technologies for national security rather than in a position of catching up to strategically important innovations and achievements of others. With no R&D facilities of its own, DARPA has become known as a

---

[107]www.darpa.mil

laboratory and incubator of innovation by providing thought leadership, community-building frameworks, technology challenges, research management, funding, and other cultural and infrastructural support elements that it takes to usher transformative ideas toward consequential new realities. DARPA's management of its portfolio reflects the fact that while the agency's mission and philosophy have held steady for decades, the world around it has changed dramatically, and the rate at which those changes have occurred has in many respects increased. Those changes include some remarkable and even astonishing scientific and technological advances that, if wisely and purposefully harnessed, have the potential not only to ensure ongoing U.S. military superiority and security, but also to catalyze societal and economic advances. At the same time, the world is experiencing some deeply disturbing technical, economic and geopolitical shifts that pose potential threats to U.S. preeminence and stability. These dueling trends of unprecedented opportunity and simmering menace, and how they can be expected to affect U.S. national security needs a decade and more from now, deeply informed this agency's most recent determination of its strategic priorities for the next several years. DARPA today is focusing its strategic investments in four main areas: rethink complex military systems; master the information explosion; harness biology as technology; and expand the technological frontier. DARPA's FY2015 budget is $2.92 billion. Some examples of basic research funding by DARPA are:

Atomic GPS

The Global Positioning System (GPS), which DARPA had an important but limited role in developing, is a great tool but maintaining it as a satellite system is increasingly expensive. DARPA does not have an explicit program to replace GPS, but its chip-scale combinatorial atomic navigation and quantum assisted sensing initiatives explore the use of atomic physics for much better sensing. If you can measure or understand how the Earth's magnetic field acceleration and position is affecting individual atoms (reduced in temperature), you can navigate without a satellite. In fact, you can achieve geo-location awareness that could be thousand times more accurate than any system currently in existence, say researchers. A drone with a quantum compass, for example, would not require satellite navigation, which would make it much easier to fly and less hackable. Future devices that understand where they are in relation to one another and their physical world will not need to rely on an expensive satellite infrastructure to work. That means having more capable and cheaper devices with geo-location capability, with the potential to improve everything from real-time, location-based searches to self-driving cars and those anticipated pizza delivery drones. The most important civilian use for quantum GPS could be privacy. Your phone would not have to get signals from space anymore to tell you where you are. It would know with atomic certainty.

The area of the electromagnetic spectrum between microwave, which we use for cell phones, and infrared, is the terahertz range. If scientists can figure out how to harness it, we could open up a vast frontier of devices that do not compete against others for spectrum access. Research into terahertz electronics has applications in the construction of so-called meta-materials, which would lend themselves to use in cloaking for jets and equipment and even perhaps invisibility. Unlike X-ray radiation, terahertz radiation is non-invasive, and hence metamaterial smart clothes made with small terahertz sensors would allow for far faster and more precise detection of chemical changes in the body, which could indicate changes in health states - the future doctor in your pocket.

A Virus Shield for the Internet of Things

The forecast is for fifty billion interconnected devices by 2020, for everything from appliances to streets, pipes and utilities through supervisory command and control systems – the Internet of Things (IoT). DARPA, through a program announced in 2012, is trying to patch the security vulnerabilities that could pervade the IoT, to prevent military vehicles, medical equipment and even drones from being hacked. In the future, some of the software tools that emerge from this program could be what keeps the civilian IoT operating safely. Its influence will be felt by what does *not* happen because of it – namely, a deadly industrial accident resulting from a catastrophic cyber-security breach. Without better security, many experts believe IoT will never reach its full potential. In order for IoT to really revolutionize the way we live it must be secure. Barriers to IoT include failure to achieve sufficient standardization and security.

Rapid Threat Assessment

The Rapid Threat Assessment program wants to speed up by orders of magnitude how quickly researchers can figure out how diseases or agents work to kill humans. Instead of months or years, DARPA wants to enable researchers to within thirty days of exposure to a human cell, map the complete molecular mechanism through which a threat agent alters cellular processes. This would give researchers the framework with which to develop medical countermeasures and mitigate threats. In the short term, this is another research area notable primarily for what does not happen after a pandemic hits. It took years and plenty of money to figure out that H5N1 bird flu became much more contagious with the presence of an amino acid in a specific position. That is what enabled it to live in mammalian lungs and, thus, potentially be spread by humans via coughing and sneezing. Knowing this secret earlier would have prevented a great many deaths. In the future, the biggest contribution of this program may be to enable future drug discovery.

Chapter 4   Innovation Research

If people really try they can figure out how to invent things that actually have an impact
— Bill Gates

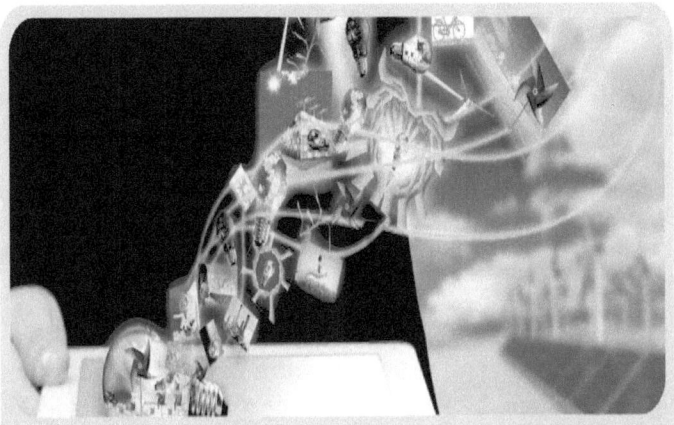

   For centuries economic thinkers have warned that technology would leave people without good work to do, only to be proved wrong. Today a surprising number of prominent economists think this time may be different. From personalized medicine to mobile payments to driverless cars, new devices and services may bring about profound declines in employment, and create an even more unequal society, divided between those who have the education and skills to take advantage of these changes and those who do not[108]. The Industrial Revolution helped overcome the limitations of muscle power to generate useful energy at will – it was the genesis of humanity's first machine age. Now the Digital Revolution - computer hardware, software, and networks - is doing for mental power what the industrial revolution did for muscle power, ushering in the second machine age. The digital world is subject to different economics with machine-to-machine communication, and zero marginal cost of reproduction making things free, perfect, and instant, bringing economic disruption with companies having less need for some kinds of workers[109].

   The nature of work in advanced economies is rapidly changing. Previous technological innovation has always delivered more long-run employment, not less, but in the next two decades, almost half of today's jobs could be automated[110]. Large firms are no longer the major providers of new jobs for Americans. Most new jobs

---

[108]Goldfarb, Zachary - Relax. The Economy Will be Just Fine, The Washington Post, February 16, 2014

[109]Brynjolfsson, Erik and McAfee, Andrew – The Second Machine Age: Work, Progress, and Prosperity in a Time of Brilliant Technologies, 2014

[110]The Economist - Briefing: The Future of Jobs; Special Report: Tech Startups, January 18, 2014

emanate from small firms that increasingly play an important role in technological change and how markets respond to this change. The highly structured and rigid organization of large firms is not conducive to undertaking risky R&D, innovative activity may most flourish in environments free of bureaucratic constraints, and larger firms reward the best researchers by promoting them out of research to management positions. In contrast, smaller firms place innovative activity at the center of their competitive strategy[111]. In the year prior to March 2013, young businesses added 2.8 million jobs, down from 4.1 million twenty years before. In 1993, more than half the jobs in the private sector were at companies employing fewer than two hundred and fifty employees. In 2013, the share of U.S. workers employed by small businesses with less than five hundred employees was about forty-seven percent.

As is the case with basic research funding explored in the previous chapter, several federal agencies also fund innovation research through the small business innovation research program. Innovation research involves translation of discoveries and inventions arising from basic research to market-leading products and services, what we can call knowledge-based innovation. This chapter provides ten examples (text box) of such research performed by startups funded by the five federal agencies described in Chapter 3 – the National Institutes of Health (innovative cancer cures; a portable photosynthesis system), the Department of Energy (a hybrid water heater using a electrochemical compressor to conserve energy; reducing photovoltaic installation cost by automating the process), The National Aeronautics and Space Administration (big data analysis on interactive maps for improved weather forecasting; a cryocooler for deep-space applications), the Defense Advanced Research Projects Agency (dramatically improving materials using nanostructures; packaging and control electronics for cryogenic, superconducting radio frequency devices), and the National Science Foundation (a wearable energy harvester that eliminates the need for batteries; production of the wonder material spider silk without spiders using synthetic biology techniques).

Federal Innovation Research Funding

In the period 1971-2006, seventy-seven out of the most important eighty-eight innovations rated by R&D Magazine's annual awards have been fully dependent on federal research support, especially, but not only, in their early phases, and these awards exclude information and communication technology innovations[112].

---

[111]Acs, Zoltan and Audretsch, David - Innovation and Small Firms, 1990

[112]Block, Fred and Keller, Matthew - Where Do Innovations Come From?, 2008

Funding by a single federal agency, the National Science Foundation, for example, has facilitated multiple technological leaps that have directly improved our daily lives. Table 4.1 provides a listing of NSF-seeded technology.

Table 4.1   Listing of NSF-seeded Technology

3D Printing
American Sign Language
Bar Codes
Biofuels
Bucky Balls
Computers and Computer Networking
Computer-aided Design, Computer-aided Manufacturing
Cellular Biology
Clean Water
Cloud Computing
Cryptography
Deep-sea Mining
DNA Sequencing and Fingerprinting
Doppler Radar
Fiber Optics
Global Positioning System
Google
Improved Accuracy of 911 calls
Improved Detection of Improvised Explosive Devices
Improved Laser Eye Surgery
Internet
Magnetic Resonance Imaging
Ozone Hole
Plant Genomics
Radio Frequency Identification
Reducing Blood Infections
Retina Chip
Routers
Screening for Counterfeit Pharmaceuticals
Search-and-Rescue Robots
Speech Recognition
Surgical Innovations
Tissue Engineering
Touchscreen Technology
Web Browser

The strong bias towards quick returns for private venture capital is damaging to the scientific exploration process which requires longer time horizons and tolerance of failure. Also, the inability of private actors to build or sustain manufacturing capabilities in various high-tech fields presents a broader problem with the nation's innovation system. The short-term approach of U.S. corporations and venture capitalists of achieving rapid financial returns often discouraged them from any interest in building domestic manufacturing capabilities while encouraging outsourcing of manufacturing as an option. Additionally, factors related to competitiveness, foreign usurpation, technology diffusion, spillovers, and intellectual property loss call for a determined policy of strategic denial. Table 4.2 shows the many products invented in the U.S. now mostly made abroad.

Table 4.2   Invented in America Now Mostly Made Abroad

Complex Robotics
e-Readers
Flat Panel Displays
Interactive Electronic Games
Laptop Computers
Liquid Crystal Display TV
Lithium Ion/Hybrid Batteries
Semiconductor Production Equipment
Semiconductors and Microprocessors
Smartphones
Solar Cells
Wind Turbines

Take the case of developing clean technologies - the key reason for uneven U.S. performance in this sector has been its heavy reliance on venture capital but venture capitalists prefer to finance technology with low capital requirements that are close to market penetration. They also lack the resources to fully finance the growth of clean technology companies that are capital intensive and competing within very complex markets. It is why virtually only public sector money is currently funding the riskiest and most capital-intensive projects in such technologies. The more recent trend indicates venture capital firms are attracted to clean technologies as a result of government support, and nearly all their funding poured into established technologies, some of which were already benefiting from decades of development. If they are not interested in capital-intensive industries, or in building factories, it is not clear what they offer in terms of economic development? The assumption that we can leave applied research to the business sector and that this will spur innovation is one with little evidence to support it.

Venture capitalists and entrepreneurs respond to government support in choosing technology to invest in, but are rarely focused on the long-term. They are also unwilling to participate in technology development that does not lead to successful initial public offerings, or mergers and acquisitions. The failures of some solar and advanced battery companies after substantial public funding highlights the "parasitic" innovation system that the U.S. created for itself, where financial interests are always and everywhere judge, jury, and executioner of all innovative investment dilemmas. Clean technologies are already teaching us that this changing world requires coordination and investment of multiple entities, otherwise R&D, support for manufacturing, and support for market creation and function will remain dead ends. Nurturing risky new industries requires support, subsidy and long-term commitments to manufacturing and markets as well. The government must be willing to support technology until they can be mass-produced and broadly deployed.

Innovative Cures for Cancer
Novel Mechanism for Targeting and Eliminating Tumor Cells

This company based in Covington, Kentucky was founded in 2006 with two employees and technology licensed from the Cincinnati Children's Hospital Medical Center. It is developing innovative cures for cancer. Unlike any other potential drug known in development, the first-in-class biologic, BXQ-350, has an entirely new and novel mechanism of action for targeting and eliminating tumor cells with potential efficacy across a broad range of tumors, while minimizing "off-target" effects. Clinical trials for BXQ-350 will include treatment of Glioblastoma multiforme (GBM), a deadly form of brain cancer. Further clinical studies may expand the market for other types of cancers, both as the primary and adjuvant therapy. The total federal funding the startup has so far received is close to $5 million, which includes a "bridge award" of nearly $3 million from the National Cancer Institute (NCI) to support testing of BXQ-350 in the clinic. The company also earned a partnership with the Nanotechnology Characterization Lab, the NCI's specific resource and knowledge base for cancer researchers with "nano" drugs. In 2015, the startup obtained FDA Orphan Drug Status for the primary active in BXQ-350 against GBM. The company has been able to leverage federal funds to attract an additional $18 million in state funds and private investment capital. The firm currently employs six people, and has multiple consultants and collaborators in toxicology, regulatory affairs, manufacturing, and clinical research. It plans to develop its products through Phase I and Phase II clinical trials, seeking additional funding options through acquisition, partnering, or licensing by major pharmaceutical companies, who will carry on Phase III clinical development, product launch, and sales and marketing. The company also has a number of out-licensing opportunities in oncology diagnostics and drug delivery.

In this context, consider the following example: Outside Detroit, a Chinese-owned company is churning out lithium ion batteries for more than 30,000 vehicles in 2015 - the company is A123 Systems, a tarnished showpiece of the federal government's stimulus-era loan program for electric vehicles. The Department of Energy awarded A123 a $249 million matching grant, one of its biggest, to build battery factories. To pay its share, A123 got about $100 million from investors including GE and Qualcomm. In 2012, after two years of losses and a damaging battery recall, A123 filed for bankruptcy protection and began looking for a buyer. The best offer came from the U.S. division of Chinese conglomerate Wanxiang Group that has been operating in the U.S. since 1994, and promised not to shut down the American operations of A123. Wanxiang paid $247 million to acquire A123 in early 2013. In 2015, revenue will reach about $300 million and the company will breakeven for the first time. Wanxiang has bought more than twenty companies including another federally-backed company,

Fisker, which was in a hurry to begin production of a $100,000 plug-in hybrid called the Karma. Amid A123's death and rebirth, the company missed out on the big U.S contracts for electric vehicles - LG supplies batteries for General Motors and Ford's electric cars, and Tesla signed with Panasonic. For now, A123's biggest orders come from China, where it supplies high-voltage batteries to carmakers Chery and Shanghai Automotive and bus maker Tianjin Santroll.

Portable Photosynthesis System
Simultaneous Analysis of Carbon Dioxide and Water Vapor

Since 2001, this small business headquartered in Lincoln, Nebraska has launched more than fifty products for cutting-edge research including drug discovery, global climate change, industrial environmental monitoring, and disease process investigation. The company first introduced scientific instruments for plant science research and quickly grew to provide scientists tools for such diverse disciplines as atmospheric research and the study of how proteins interact at the cellular level. It developed the first commercially viable portable photosynthesis system allowing scientists to obtain simultaneous conductance and photosynthetic values in the field. In 1967, the Rockefeller Foundation funded a large grant to develop grain sorghum as a food supplement for use in under-developed countries. A group of scientists working on the project at the University of Nebraska, Lincoln (UNL) hired the founder to develop specialized instrumentation to help their agricultural research. Shortly after a Lincoln-based symposium on this project ended, the UNL group began to receive requests for the instrumentation developed for the project from many of the scientists who had attended. While working on the sorghum project, the founder decided to pursue a graduate degree in engineering. His thesis consisted of developing a light meter, much of which was later incorporated into the LI-185 quantum/radiometer/photometer that the firm first marketed in 1971. That same year, the founder set up a company to manufacture these sensors and other products. The company received its first SBIR award in 1994 when it had 98 employees.

One example of the company's success through SBIR relates to a Phase I grant to develop a new open-path carbon dioxide analyzer for measuring gas fluxes. Two years after receiving this award the firm released the analyzer that could measure both CO_2 and water vapor simultaneously at high speed with precision and accuracy. The instrument has become a standard used worldwide. It is estimated that more than 80% of the measurements examining carbon balance of agricultural and natural ecosystems have been made with this company's instruments. This small business is a global leader in the design, manufacture, and marketing of high quality, innovative instruments, software, reagents, and integrated systems for plant biology, biotechnology, drug discovery, and environmental research. Over 30,000 customers in more than 100 countries use its instruments. The company also sells products through a global network of distributors. The firm has grown to include more than 300 employees at its headquarters in Lincoln and at its subsidiaries in Bad Homburg, Germany and Cambridge, UK. It is a privately held company and is ISO 9001:2008 certified. In all, the firm has received 26 awards, 19 from the National Institutes of Health totaling $4.2 million and 7 from the Department of Energy for a total of $2.9 million.

Entrepreneurship is the main engine for economies to evolve and regenerate. Adversity, like necessity, breeds inventiveness. It is not just talent but tenacity, insatiable questioning of authority, determined informality, combined with teamwork, mission, risk, cross-disciplinary creativity, and a unique attitude toward failure[113].

> *Hybrid Water Heater*
> *Electrochemical Compressor for 50-Gallon Residential Hot Water System*
>
> This startup based in Seaford, Delaware received SBIR grants from the Department of Energy (DOE) totaling $1.2M to address the problem of existing extensive use of energy-hungry hot water heaters in both residential and commercial building in the United States. In addition, attempts to create a low energy hybrid hot water heater have resulted in hot water heaters so loud that they cannot feasibly be used in residential applications. This R&D work entails the development and prototyping of a hybrid hot water heater compressor which drastically reduces the energy needed to heat hot water as well as reducing the sound emitted from the heater. The first phase successfully created a new electrochemical compressor suitable for a 50-gallon hybrid hot water system that is more efficient than current mechanical compressors and that is also noiseless, vibration free, modular and scalable utilizing water as a refrigerant. The goal of the next phase will be to create a high fidelity heat pump hybrid hot water heater in conjunction with General Electric allowing this unit to determine how to meet DOE performance and cost targets for the commercial and residential electrochemical based hybrid hot water heater. More than 70% of electric demand is utilized in buildings, with approximately 15% of residential electric demand being for hot water production. Anything that will reduce that individual subunit demand will significantly reduce our total energy usage, emissions from producing electricity, and our dependence on foreign energy sources. In addition to energy conservation, this product would also reduce the use of greenhouse gas refrigerants, resulting in substantial environmental benefits. This is a disruptive technology that will also allow this compressor technology to be modified to other applications such as wine chillers and electronic cooling.

Entrepreneurship is the central comparative advantage for when entrepreneurs succeed, they revolutionize markets. Even when they fail, they keep incumbents under constant competitive pressure and thus stimulate progress. A startup is dedicated to creating something new under conditions of extreme uncertainty, and where execution might be of greater importance than strategy. Founders of new businesses, who face significant capital constraints and great risk, rely on opportunistic adaptation to unexpected events. Only 12% of the founders attributed the success of their companies to "an unusual or extraordinary idea", the rest reported their success was mainly due to the "exceptional execution of an ordinary idea"[114]. Most startups fail, but many of those failures are preventable. The main bottleneck on innovation is the time it takes society to sort through the many combinations and permutations of new technologies and business models. The lean startup approach fosters companies that are both

---

[113]Senor, Dan and Singer, Saul - Start-up Nation: The Story of Israel's Economic Miracle, 2011

[114]Bhide, Amar – The Origin and Evolution of New Businesses, 2000

---

more capital efficient and that leverage human creativity more effectively. Inspired by lessons from lean manufacturing, it relies on validated learning, rapid scientific experimentation, as well as a

---

Reduce Cost of Solar Energy
Automated Utility-scale Photovoltaic Panel Installation System

With solar power being seen as of national strategic importance, the DOE established the SunShot Initiative to drive an aggressive goal of $1/watt peak (Wp) installed price from the current cost of $2.34/Wp for large, utility- scale photovoltaic (PV) installations. Watt peak capacity is the nominal capacity, or the capacity which is realized under certain test conditions; it is not the regular power output, but instead the maximum capacity of a module under optimal conditions. As PV module prices have dropped precipitously due to global oversupply, Balance of System (BOS) or non-module costs such as installation labor, structural materials, and site preparation have come down much slower to the point that they now make up a majority of the overall installation expense. A typical 10 megawatt-direct current plant will require 30,000 – 40,000 modules on 12.5 -17.0 kilometers of racking; currently each step of the installation process is carried out by hand, being both time consuming and labor intensive. To specifically address this problem, this startup based in San Jose, California was provided grants totaling $850,000 to help dramatically reduce BOS costs through the introduction of automation and robotics into the repetitive steps of the large-scale PV installation process.

The startup's system consists of three components: the Panel Assembly Cell (PAC), a mobile factory where individual modules are preassembled and wired into larger panels, the WaveRack ground mount, a simplified fixed tilt ground mount support system that doubles as an above ground transport track for panel installation, and the PV autoloader/shuttle, an automated system that takes preassembled panels from a centralized logistics area and installs them at megawatts per day rates. This system can dramatically reduce labor, heavy equipment, and overhead costs as well as construction times. In the first phase the company developed the PV autoloader to supplement the prototype PAC, WaveRack and PV shuttle sub-systems. In the next phase, the company seeks to fully productize their system by taking all three components from alpha to full production.

---

number of counter-intuitive practices that shorten product development cycles, measure actual progress without resorting to vanity metrics, and learn what customers really want. It enables a company to shift directions with agility, and offers a way to test their vision continuously, to adapt and adjust before it is too late, by providing a scientific approach to creating and managing successful startups in an age when companies need to innovate more than ever[115].

The small business innovation research program was first conceived at the National Science Foundation. In 1976, Roland Tibbetts, a program officer at this agency initiated a pilot program to support small business, specifically to provide early-stage financial support for high-risk technologies with commercial promise. Based on

---

[115]Reis, Eric - The Lean Startup: How Today's Entrepreneurs Use Continuous Innovation to Create Radically Successful Businesses, 2011

this pilot, the U.S. Congress passed the Small Business Innovation Development Act of 1982. The SBIR program was established under this Act with the purpose of strengthening the role of innovative small businesses in federally-funded R&D. Each year, federal agencies with extramural R&D budgets exceeding $100 million are required to allocate a fixed percentage, historically 2.65%, increasing in yearly steps to 3.5% in 2017, of their R&D budget to this program. Today, eleven federal agencies independently offer and administer this program. Table 4.3 provides data on the overall federal SBIR program as of January 2015[116].

Table 4.3   SBIR by the Numbers

Total Awardees	22,084
Currently Active Firms	6,045
Scientists and Engineers Employed	400,000
Patents Issued	120,924
Venture Capital-funded Firms	2,723
Mergers & Acquisitions	1,866
Total Phase I Awards	105,208
Total Phase II Awards	41,782
Total Awards	146,990
Annual Budget	$2.8 Billion
Total Federal Investment	**$40.8 Billion**

This program targets the entrepreneurial sector because that is where most innovators and their innovations thrive. However, the risk and expense of conducting high-tech R&D are often beyond the means of many small businesses. By reserving a specific percentage of federal R&D funds for small businesses, the SBIR program protects and enables them to compete on the same level as larger businesses. It funds the critical startup and development stages to encourage the commercialization of the technology, product, or service, which, in turn, stimulates the U.S. economy. Table 4.4 features four of numerous highly successful companies that were seeded by this program. Such small business contributions have enhanced our nation's defense, protected our environment, advanced healthcare, and improved our ability to manage information and process data.

The program was reauthorized until 2008 by the Small Business Reauthorization Act of 2000. Subsequently, Congress passed numerous extensions, the most recent of which extends the program through 2017. The SBIR program invests federal research funds to support scientific excellence and technological innovation to help build a strong U.S. economy. The goals of this program are to

---

[116]Eskesen, Anne - Innovation Development Institute, 2013

stimulate technological innovation; meet federal research and development needs; foster and encourage participation in innovation and entrepreneurship by socially and economically disadvantaged

Table 4.4   NSF SBIR-seeded Fantastic Four

Company	Founded	Description	Market Cap
ecoATM	2008	Automates inspection, pricing, payment, and take-back of used consumer electronics; 270 employees, 1100+ kiosks in 42 states; acquired by Outerwall in 2013	$350M
Intralase	1997	California-based developer of femtosecond laser technology, acquired in 2007 by Advanced Medical Optics, a maker of medical devices and laser vision-correction systems	$808M
Qualcomm	1985	Andrew Viterbi and six colleagues formed 'QUALity COMMunications'; in 1987–88 NSF SBIR provided $265,000 for single chip implementation of Viterbi decoder; led to high-speed data transmission via wireless and satellite; holds more than 10,100 U.S. patents, licensed to more than 165 companies	$78B
Symantec	1981	Gary Hendrix founded Symantec; in 1982, NSF SBIR awarded $30,000 to develop a framework for managing dissimilar data; now makes security, storage, and backup software; most used certification authority	$12B

persons; and increase private-sector commercialization of innovations derived from federal research and development funding. It is a highly competitive grant- and contract-based source of funding for domestic small businesses to encourage them to engage in federal R&D that has commercial potential. It enables such business entities to explore their technological potential while providing incentives to profit from its commercialization. This stimulates high-tech innovation and the nation gains entrepreneurial spirit while simultaneously fulfilling the country's R&D needs.

The quasi-public good nature of several technologies funded by this program can be disaggregated into the following technology elements:

- New technology platforms, proofs-of-concept, or generic technologies
- Infra-technologies and associated standards

- Proprietary technologies

Infra-technologies provide the basis for technical and functional interfaces between products and components that make up a system. It requires interoperability protocols to share data with multiple tiers in the supply-chain, and the measurement and test methods required to efficiently conduct R&D, control production, and

---

Maps, Data, and People
Big Data Analysis on Interactive Maps

This small company based in Halethorpe, Maryland was provided federal R&D funding to develop a product based on emerging map and geographic information systems that allow for big data analysis on interactive maps. This firm developed a patent-pending, collaborative geospatial intelligence solution to aid public safety officials to interact with maps, data, and people simultaneously. This small business has evolved from a specialized weather and climate technology communications entity to a cyber and homeland security technology development and applications company using its expertise in collaborative cross-platform virtual globe collaboration, visualization of earth science data using 3-D virtual globes, and real-time collaboration using geobrowsers. This firm sees a broad spectrum of applications for its product in fields such as homeland security, weather forecasting, transportation, energy management, and education. The need for real-time, actionable information is critical during day-to-day and emergency response operations where multiple jurisdictions and disciplines interact. Plenty of homeland security-related information exists at the local, tribal, state, and federal levels, but since equipment investment decisions have been made based on the specific operational needs of individual agencies without benefit of a national strategy or standards, this information is often trapped in silos. As a result, potentially critical information often does not make it into the hands of people who need it the most.

Extreme weather events are on the rise, climate change is impacting regions and nations around the world and geo-hazards such as volcanoes and earthquakes have had a tremendous economic and psychological impact on significant world populations. This startup's technology is addressing the need to perform real-time visualization and collaboration sessions across multiple agencies and organizations in order to accelerate and crystallize situational awareness for enhanced decision making. Whether it is tracking critical cargo, people or assets collaboratively or positioning relief crews and supplies, its product enables real-time sharing of data between public and private agencies for enhanced decision making for the protection of life and property.

---

execute marketplace transactions. Achieving efficiency for technologically advanced production systems require process-control techniques, critical evaluation of engineering data to execute the process control, and methods and reference materials to calibrate complex equipment. Product acceptance testing protocols and standards and specialized facilities for determining compliance with industry standards are efforts to develop the industrial commons for various future technologies. More knowledge of technology life-cycles show that various elements of emerging technologies often advance at very different rates, leading to incomplete technology platforms and thus lower rates of innovation.

The driving premise of the SBIR program is to execute a high-risk, high-reward model by facilitating private-public partnerships to help seed whole new industries in America. The program plays a pro-active role in accelerating the innovation process by increasing industry participation in university research, and stimulating entrepreneurship

---

Low-Temperature Device for Deep Space
High-frequency Single-stage Pulse Tube Cryocooler

This firm based in Madison, Wisconsin was awarded a grant worth $875,000 from NASA to develop a high-frequency, low-temperature pulse-tube cryocooler that is an enabling technology for far-flung future NASA missions that have the need to preserve mission propellants in their liquefied state until needed to maneuver near a destination or a midway point. Such missions will require cooling capacities in excess of 0.3 watts at 35 degrees kelvin with heat rejection capability to temperature sinks as low as 150 kelvin (K) at input powers up to 20 watts. Presently there are no cooling systems operating at such low temperatures. The innovation is a high-frequency single-stage pulse tube cryocooler that operates at a heat rejection temperature of 150K. It employs a flexure-bearing cold compressor also operating at 150K. The unique built-in design features of the proposed unit will avoid all thermal expansion issues enabling it to operate within a cold, 150K environment. This cooler addresses the need to prevent boil-off of cryogenic propellants on long duration remote missions. Due to its high heat capacity regenerator matrix, the cooler has a high efficiency and a small footprint, making its launch mass minimal. The coherent regenerator matrix configuration prevents movement and so prevents degradation over time. Due to its unique compressor being cold-tolerant down to 150K the cooler can continue to run even where the sun is dim and far away that the energy captured by the spacecraft cannot keep the compressor warm. The compressor being designed to run cold and on minimal input power allows it to operate at the inherently low solar intensity, which seriously depresses solar cell power generation.

The design concept calls for using regenerator materials from a recently developed class of high heat capacity rare earth alloys; a compressor and cold-head design optimized for a low temperature heat sink; and minimizing the known losses in the pulse tube proper. Pertinent NASA projects include missions to the outer planets and their moons, such as Europa, Titan, Neptune, and Triton. These missions need to carry considerable propellants for many years to be used in braking and/or orbital insertion maneuvers at the target location(s). High thrust propulsion is required for these near planetary maneuvers. Significant benefit can be gained from improved mission performance, which is greatly impacted by propulsion system mass. Propellant is usually the predominant contributor to the mass of such chemical systems. Therefore, the largest potential system and mission performance gains are likely from improved specific impulse derived from the use of cryogenic propellants.

---

and commercialization. The program is mindful of its unique character and function, and how these influence its strategic approach. The preponderance of federally-funded basic research supports the discovery that enables innovation - it is discovery *for* innovation. It lays the scientific and engineering knowledge base for technological innovation to prosper. SBIR-funded research and development supports both industry and academia to ignite the translation of discovery to market, to realize prototypes leading to market-ready products and services - it is discovery *to* innovation. It invests in cutting-edge, high-risk, high-quality R&D in science, engineering, and

---

*Dramatically Improve Traditional Materials
Technology Based on Nanostructured Chemicals*

This company based in Hattiesburg, Mississippi develops and commercializes its innovative polyhedral oliogomeric silsesquioxane (POSS) nanotechnology based on nanostructured chemicals. It has so far obtained 20 SBIR awards totaling $5 million, out of which $3.2 million was from the Department of Defense. The technology and company were originally spun out of the U.S. Air Force Research Laboratory and has grown from six employees to 20 employees currently. POSS is a revolutionary new chemical based on silicon-derived building blocks that provide nanometer-scale control to dramatically improve the properties of traditional polymers, colorants and filters. In addition to being biocompatible and recyclable, POSS releases no volatile organic compounds which can have negative environmental effects. In 2005, a Presidential Determination deemed POSS nanotechnology to be in the strategic national interest of the United States. This allowed the company to secure facility funding through Title III of the Defense Production Act. The act is to ensure that strategically important technologies continued to be manufactured in the United States. The Title III effort is coordinated by the Air Force from the Wright Patterson Air Force Base in Dayton, OH. The company's technology has commercial applications in areas ranging from biomedical, such as synthetic organ replacement, to electronics, such as next-generation microchips with reduced feature sizes; from aviation, such as new filter filaments and composites, to rubbers for improved downhole oil and gas technology capabilities. A specific example is provided by an $850,000 grant to develop a low-cost and versatile method for shielding commercial and military single crystal solar cells from damage against proton and electron radiation. The firm also developed a related coating variant for the protection of thin film photovoltaics against atomic oxygen, vacuum ultraviolet radiation and improved voltage stand-off. The technical approach utilizes metallized nanoscopic POSS as conformal coatings on solar cell surfaces. Such coatings would permit spacecraft designers to increase duty cycles while operating in half-geo orbits. For this purpose, the firm also conducted fundamental radiation and environmental testing to determine shielding effectiveness in addition to developing processes to enable low-cost spray application of the coatings.

---

education. It aims to systematically facilitate and accelerate the use of basic research results to impact U.S. innovation capability. The investment goal is to promote innovation that benefits society and the nation through successful commercialization. SBIR grants are designed to support small businesses at the earliest investment stage when technical risk is highest and when financial and other support could possibly aid in driving transformational ideas and research to the marketplace. Support to businesses takes two forms: direct investment in small businesses for R&D and to establish partnerships, and indirect incentives for large businesses to collaborate with academia and small business.

Over the years, various agencies have instituted a number of supplemental opportunities available to its startups during their Phase II effort. One such opportunity is the Phase IIB funding mechanism initiated by the NSF SBIR program in 1998, and the recent focus of a study by the National Academies of Sciences, Engineering, and Medicine that provided positive testimony on it to the U.S. Congress.

Cryogenic Radio Frequency Technology
Device Packaging and Control Electronics

This small business based in San Francisco, California has grown from two employees when it received its first SBIR award from the Department of Defense (DOD) in 2005 to 9 employees today. It has so far received 15 grants totaling $6.8 million, mostly from DOD. The U.S. Navy purchased from this firm over 60 radio frequency (RF) distribution units under a $10 million production contract. The company's business strategy is to provide superior technology to the military and intelligence end-user customers exclusively. In one example, the company was awarded an additional development contract with the Space and Naval Warfare Systems Command, and leveraged that work to win a Rapid Innovation Fund contract worth $1.5 million through the Office of Naval Research. These contracts resulted in a full prototype delivered to the Navy, which in turn led to a contract with a major prime contractor – the company has delivered four low-rate initial production units and is in the process of proposing a 40-unit production order. This small firm responded to the Navy's need for effective electromagnetic interference mitigation to reduce co-located interference, common battle group interference, and jamming signals by developing and transitioning high-performance cryogenic and superconducting RF components. It not only developed this cutting-edge technology but also performed testing to demonstrate full military qualification in harsh environments. Another example is provided by a grant to analyze and design a physically small superconducting antenna for high-frequency direction finding with a set of target specifications. The company modeled and simulated a superconducting antenna for this application, including the control electronics, and completed an array design and device fabrication. It then measured noise properties of this superconducting quantum interference device array. Commercial applications include satellite communications and biomagnetic detectors for magnetocardiography.

This supplement funding provides additional matching dollars predicated on a startup's ability to obtain private third-party investment. It was recognized early on that addressing commercial and financial issues throughout the initial stages of technology development, concurrently gathering information about markets, potential customers, competitors, strategic direction, and finance, are critical to effective and timely commercialization. The program is aggressive in requiring early attention to business issues.

The Technology Enhancement for Commercial Partnerships is an additional supplement opportunity that is specifically designed to increase startup technology commercialization success rates by helping startups co-develop their technology with strategic industrial partners. This is recognized as a critical success factor to move technologies developed by startups to market. Potential partners, however, frequently demand technical specifications and require proof-of-concept data as a prerequisite for partnering that is often beyond the scope of the project objectives of the grant. This supplemental funding enables federally-funded startups to conduct additional research to successfully meet the requirements of a corporate partner leading to marketable products and services. This opportunity is intended to challenge startups to begin to develop an outward focus, and to more

---

Spider Silk without Spiders
Production of Synthetic Spider Silk Fibers

This National Science Foundation SBIR-funded research and development work from March 2012 to February 2016 in the amount of approximately $1.0 million to a startup based in San Francisco, California was to develop and commercialize synthetic spider silk fibers. Spider silk is a unique material in nature that is currently inaccessible on a commercial scale. Spider silk and other protein polymers are broadly useful in fields ranging from specialty textiles to medical devices and advanced composites. The critical limitation in producing artificial spider silk fibers has been the lack of availability of bulk silk material and the knowledge of how to appropriately process the polymer into a product of native quality. This R&D effort will deliver scalable quantities of material through microbial production of spider silk protein using a commercially viable cost structure. In addition, it will examine the key parameters for processing silk polymer into fibers whose properties surpass those of native spider silk. The ability to produce prototype silk fibers from recombinant protein will enable the initial steps towards commercializing spider silk fiber-based products. Spiders, like other insects, have the ability to produce natural silk fibers with useful properties such as high tensile strength, durability, softness, and elasticity. This small business has developed new technology to replicate this natural silk production process sustainably on a large scale. The startup studies the silk proteins in nature to understand what gives them their unique properties. It then develops proteins inspired by these natural silks by placing genes in yeast, and producing them in large quantities through a fermentation process using yeast, sugar, and water. The company then spins the silk protein into fibers to knit or weave into fabrics and garments. The broader impact is the adoption of a job-creating bio-based economy in the United States. The ability to produce protein polymers has bedeviled biological researchers for decades. Many important structural proteins and enticing commercially-useful materials have remained effectively impossible to produce. The advent of cutting-edge techniques in synthetic biology, microfabrication, and materials processing now make the production of protein polymers and the processing of them into beneficial products a realistic goal. Potential applications of protein-based polymers include a full range of sophisticated materials that are also "green" and sustainable. Spider silk polymers, due to their mechanical properties, can potentially be used to create the next generation of ballistic fibers in the production of armor for military, law enforcement, and private users. In addition, the ability to produce advanced polymers independent of petroleum sources is a key goal of the emerging bio-based economy. Many protein polymers, including silk, are biocompatible and biodegradable and thus can form the basis for new classes of medical materials used to replace or re-grow connective tissues with implants or devices.

---

rigorously evaluate their strategic business and commercialization options. This additional research not only benefits the startup but also provides a mechanism for large and mid-sized corporations to provide input into the commercial development of new technologies, products and services.

There is now a demonstrated negative trend in early-stage private sector investment, with venture capital firms often not considering or even abandoning ideas at this crucial stage. Large industries, on the other hand, have moved away from early-stage

---

No Batteries Required
Wearable Energy Harvester and Wireless Sensor Networks

This startup based in Corvallis, Oregon was provided federal funding from April 2011 to March 2015 in the amount of approximately $1.0 million for R&D work to address the need for renewably-powered always-available energy for personal medical and other location-aware sensors. The R&D effort involved wearable thermoelectric generator technology yielding new advances in miniaturization, increases in power densities, application of advanced heat transfer materials, and integration with cutting-edge locator system electronics. The research focused on matching thermal resistances of the thermoelectric generator and the skin-to-air interface, accomplished by optimizing thermocouple geometries implemented in thin film semiconductors applied to a flexible substrate. The result is a fully functional wristband locator system that is lightweight, adjustable, waterproof, and renewably powered by the human body. The broader impact includes services for elderly home healthcare, location tracking of people stricken by Alzheimer's disease, and nursing home patients. As the population ages, achieving a balance between personal independence while providing primary healthcare monitoring will be critical. Wearable thermoelectric generator technology can be used to power wireless sensors that monitor patient location and help facilities track "at-risk" residents. Additionally, wireless sensors can help healthcare providers improve treatment, increase efficiency, and cut costs. Other follow-on medical applications include glucose monitoring for diabetic treatment and care, diagnosing sleep disorders, and the physiological monitoring of first responders and law enforcement personnel limited by finite and limited battery life.

Harvesting body heat and converting to usable electrical energy advances the era of autonomous wearable devices. The essential idea is to power wireless sensors and other low-power mobile electronic devices utilizing body heat as a constantly available power source. This product is a unique thermoelectric energy harvesting technology designed to power body-worn electronic devices. Integrated into wearable form-factors, such as wristbands or clothing, this product absorbs heat from the body, which is converted into electrical energy that serves as an always-available renewable power source. Similar to sunlight exciting electrons in a solar cell, body-heat absorbed this way excites electrons and optimizes this energy for body-worn medical, fitness, and safety related electronics. In the wearables market, commercial volume increments will be in the millions - in fact, a million may be a relatively small run. In this market segment, the company plans to use a component-supplier model and, because of the large volumes involved, is assessing foundries around the world. There are significant benefits to partnering with high-volume manufacturers as strategic partners for wearables. The result of decades of development, this unique flexible thermoelectric material is being incorporated into strategic partner products. The startup has world-class research partners such as the Pacific Northwest National Laboratory, the Oregon Nanoscience and Microtechnologies Institute, the Center for Advanced Materials Characterization in Oregon, and the University of Oregon.

---

research and are seeking a higher degree of technology development fidelity for potential insertion into their firms. Furthermore, there is also the immediate need to confront global competitive threats and forestall the appropriation of U.S. innovations.

Chapter 5   Infrastructure

If you do not expect the unexpected you will not find it, for it is not to be reached by search or trial
— Heraclitus

Innovation research, as discussed in the previous chapter, aims to transform public investments in scientific research to marketable products and services in an effort to seed twenty-first century industries in America. It purports to be a high-tech, high-wage, sustainable American job-creation engine. It is from such a perspective that this chapter describes the components of an innovation ecosystem. Human and knowledge capital encompasses researchers, innovators, entrepreneurs, mentors and facilitators, technical and business advisors, lead-users, and venturesome customers. In the evolution of a startup, it is, above all, the business model that is the key to a viable and successful small firm. It is what helps bring key players together for cross-pollination and triumph. It may shift as inputs from lead-users and potential customers in the target market are gathered and understood. We consider ten specific pillars and partners as they pertain to technology-intensive startups funded by the SBIR program. This chapter profiles ten startups to illustrate each of the ten pillars and partners.

Pillars and Partners

The pillars and partners that form the innovation ecosystem can be visualized as the ten spokes of a hub-and-spokes wheel of innovation with the startup at the hub. These spokes are the entrepreneurs; intellectual property rights; early adopters; technology-specific networks; industry; manufacturing partners; investors; universities; government; and support networks. These are described in the following sections.

Entrepreneurs

An entrepreneur, artist of the marketplace, continually in search of creative new commercial narratives that can capture an audience, tells a compelling story, and brings people into the world he/she has crafted. Entrepreneurs undertake personal economic risk to create an organization to exploit a new technology or innovative process that generates value to others. They see their efforts in creative terms as more art than industry; far more driven by the creative act than the almighty dollar. It is this entrepreneurial spirit that drives them to innovate, reduce marginal costs, bring cheaper products and services to the market, and spur economic growth. It is by being continuously mindful of the needs, desires, and wants of buyers and servicing them that capitalist entrepreneurs thrive; an entrepreneur or firm that is not looking out for the welfare of prospective customers is not going to stay in business for very long.

Entrepreneurship is a vital element of the U.S. economy – 80% of American millionaires are self-made, and 15% of the U.S. population runs their own companies. Freedom of self-determination spawns an extraordinary culture of work. This work ethic engendered by individual freedom and social mobility is the fundamental driver of America's economic evolution. The U.S. has a high rate of new business starts. It breeds a constant flow of such high-impact firms, the kind that creates value and stimulates growth by bringing new ideas to market, new technologies, new business models, or new and better ways of performing routine tasks. America's comparative advantage is that it has evolved a multifaceted system for nurturing entrepreneurship. Like most successful entrepreneurial ecosystems, it has emerged from the bottom-up, propelled by market forces[117].

Entrepreneurship is the act that endows resources with a new capacity to create wealth. By its very nature it is, like technology, an activity that feeds on itself. The test of an innovation lies not in its novelty, its scientific content, or its cleverness rather it lies in its success in the marketplace. Discomfort, and it may be intellectual discomfort, is the source of all entrepreneurial activity. Capitalism would slowly wither through systematization, giving way to bureaucratic capitalism. The entrepreneurial spirit, motor of capitalistic progress, would be stilled by process and routine[118]. Entrepreneurs will have to learn to practice systematic innovation, the purposeful and organized search for changes, and the analysis of the opportunities such changes might offer for economic or social innovation[119].

---

[117]Note: For each chapter subsection describing a spoke, a startup example illustrates the particular spoke.

[118]Schumpeter, Joseph – Capitalism, Socialism and Democracy, 1942

[119]Schramm, Carl - The Entrepreneurial Imperative: How America's Economic Miracle Will Reshape the World (and Change Your Life), 2006

The 1980s saw a profound shift from a managerial to an entrepreneurial economy in America. In previous eras, technology innovations and entrepreneurship drove changes and progress in bursts; now primarily because of technology advancements, innovation is smoother and we enjoy high rates of entrepreneurship. Another hallmark of an entrepreneurial economy is the constant creation and

---

Affordable High-speed Internet
Low-cost Transparent Wireless Mesh Network Node

The founder of this startup is a serial entrepreneur, this being his fourth enterprise. The publicly-funded R&D work from April 2013 to March 2016 in the amount of approximately $868,000 to this small business based in Boston, Massachusetts is to advance antenna technology and wireless network performance by developing a new kind of wireless mesh network node consisting of an integrated router, radio and antenna; and the software necessary to create a network of such devices. Dramatic capacity gains in urban wireless networks are achieved by deploying directional antennas to minimize interference and allow substantial frequency reuse. Indoor mounting of nodes in windows is desirable to avoid complications with landlords, avoid weather problems, and decrease radio frequency (RF) path loss. Finally, larger antennas produce more directionality, so the desired wireless node needs a large surface area and yet must be something that people are willing to hang in their windows, i.e. transparent and attractive. This project is using transparent conductive coatings with a novel combination of beam-steering and multiple input multiple output radio technology to develop a mostly transparent wireless node that achieves a less than15 degree wide beam that is electronically steerable over greater than110 degrees. The impact would be increased U.S. competitiveness by facilitating the spread of high-speed Internet access and through the sale of innovative new devices around the world. Community networks will deploy these nodes, as they are the lowest cost and simplest way to create an inter-building network. In particular, delivering signals via windows overcomes RF absorption in the masonry structures typical of public housing projects. Successful commercialization of this technology will significantly reduce the cost and increase the performance of urban wireless networks, thus helping extend Internet access to unserved and under-served neighborhoods and reducing the cost of offloading 3G/4G wireless traffic to local Wi-Fi hotspots. The startup provides affordable, high-speed internet access to residences and businesses as an alternative to the two large incumbents that force people into high-cost service bundles they do not want. The company offers two types of broadband internet service: "dedicated circuits" which are business-grade data lines with committed information rates, and "internet service", which are like cable modem or digital subscriber lines. Most companies prefer the consistency and reliability of dedicated circuits, but for small businesses or lower priority traffic, internet service may be sufficient. Businesses connect by installing two or three directional 802.11n radios connected to a mini-router on their rooftop to create a "node" in the startup's network.

---

destruction of jobs and businesses with the overall result being positive growth and constant renewal. Entrepreneurs now create more than half the new jobs in America. Encouraging, managing, and supporting an entrepreneurial economy is central to our place in the world going forward[120].

---

[120]Baumol, William - The Micro-theory of Innovative Entrepreneurship (The Kauffman Foundation Series on Innovation and Entrepreneurship), 2010

Four sources of innovative opportunity lie within the enterprise: The unexpected; the incongruity; innovation based on process need; and changes in industry or market structure. Another three involve changes outside the enterprise: Demographics; changes in perception, mood, and meaning; and new knowledge[121]. Knowledge-based innovation, the kind that is supported by the SBIR program, is based on the convergence of several different kinds of knowledge, not all of them scientific or technological. A knowledge-based innovator needs to learn and practice entrepreneurial management which is more crucial to this kind of innovation than to any other kind. Its risks are high thus requiring a much higher premium on foresight, both financial and managerial, and on being market-focused and market-driven. These types of entrepreneurs tend to be infatuated with their own technology, often believing that "quality" means what is technically sophisticated rather than what gives value to the user. There is a window of a few years during which a new venture must establish itself in any new knowledge-based industry. Science- and technology-based innovators find time working against them. Knowledge-based innovation and in particular high-tech innovation is inherently risky and has the longest lead-time between investment and profitability.

The focus on profits is the wrong focus for a new venture, for cash flow, capital, and controls come much earlier. Growth has to be fed. It requires cash flow analysis, cash flow forecasts, and cash management. Entrepreneurship demands financial management. A growing new venture should know twelve months ahead of time how much cash it will need, when, and for what purpose. Often it is when the opportunities are the greatest that such a venture is under severe cash pressure. Entrepreneurial strategies, purposeful innovation, and entrepreneurial management – the three together make up innovation and entrepreneurship. Entrepreneurial strategy remains the decision-making area of entrepreneurship and therefore the risk-taking one, it is by no means hunch or gamble, but it is also not precisely science, rather it is judgment. Innovation and entrepreneurship are needed in society as much as in the economy. What we need is an entrepreneurial society in which innovation and entrepreneurship are normal, steady, and continuous. There must be an economy full of innovators and entrepreneurs, with entrepreneurial vision and entrepreneurial values, with access to capital, and filled with entrepreneurial vigor. In terms of job creation, high-tech is the maker of tomorrow rather than the maker of today.

Intellectual Property Rights

Intellectual Property (IP) is a property like no other. Economists know that intangibles such as ideas, knowledge,

---

[121]Drucker, Peter - Innovation and Entrepreneurship, 2006

technology, and scientific discoveries drive the process of growth. The problems of defining and measuring the magnitude of inventions; the relationship between the technological and economic significance of the invention; the distinction between the cost of producing the invention and the value it creates; and the consequences of the highly skewed distribution of the value of inventions have been previously considered[122]. Patents do not actually provide an affirmative right to market an invention; they provide only a right to exclude others from doing so. Patents are critical for R&D investment in the pharmaceutical industry. In most other industries, they do not pose much of a barrier to imitation, and firms rely on other means, such as lead-time advantages and trade secrecy, to obtain returns on their R&D investments. Chemical patents are much more valuable, and are much less likely to be litigated. Patents on chemicals work particularly well because these patents have very well-defined boundaries whereas complex technologies such as electronics and computers might have relatively poorly defined patent boundaries. Software patents often have boundaries that are especially difficult to determine[123].

Small firms generally receive benefits that exceed costs but the net incentives for these patentees are not large. Most recent major inventions originated in large organizations although a significant number of important inventions are developed by independent inventors or inventors working in small firms. The patent system works for small inventors but does so only weakly, in part because small inventors do not have access to the resources needed to commercialize inventions; they cannot quickly ramp up manufacturing and marketing, they do not have established distribution channels, and they cannot easily finance acquisition of these assets. For small high-tech firms the high cost of getting and enforcing patents often leads them to choose trade-secret protection instead of patent protection. They use a mix of strategies to appropriate value from their inventions. The purchaser of a complex technology will often need to purchase more than just the patent, some critical knowledge might not be patented but kept instead as a trade secret and other knowledge might be tacit[124].

U.S. firms spend more than twice as much on patent and other litigation as they do on R&D. The most important cause of gridlock in R&D is not any particular lawsuit or even uncertainty about patent scope or validity but the sheer multiplicity of rights that must be acquired to undertake innovation of any sort, including drug development, gene therapy, agricultural research, semiconductors,

[122]Kuznets, Simon - Inventive Activity: Problems of Definition and Measurement, 1962

[123]Jaffe, Adam et al - Patents, Citations, and Innovations: A Window on the Knowledge Economy, 2005

[124]Bessen, James and Meurer, Michael - Patent Failure: How Judges, Bureaucrats, and Lawyers put Innovators at Risk, 2009

information and communication technologies, research tools, spectrum usage, software, nanotechnology, and even academic research[125].

One-transistor Memory Device
Chips For High-performance, Low-power Applications

This startup based in Cupertino, California is being provided federal funding from September 2012 to May 2016 in the amount of approximately $842,000 for R&D work to develop a novel one-transistor memory device, which has both volatile and non-volatile functionality, through collaboration with a commercial foundry partner. This device combines non-volatile memory's ability to retain information in the absence of power, such as flash memory, and the fast access speed and reliability of volatile memory such as static random access memory. In addition to the memory cell development and optimization, a memory block will also be developed to help commercialize this technology. This effort, if successful, will enable power-efficient computing applications and smaller form-factors for mobile devices. It can, for example, be used to reduce power consumption in data centers. Data centers' annual energy consumption is estimated to be 150 billion kilowatt-hours. A power-efficient memory such as the one being developed can reduce the overall data centers' power consumption by up to 75%. Another application is to provide an integrated memory solution - many electronic devices currently employ multiple types of memory with their own distinct characteristics. The proposed device will be able to combine different types of memory into a single memory device, simplifying the manufacturing process for embedded memory.

This startup is a semiconductor intellectual property company that develops and licenses novel 1-transistor memory technologies that provide more than 80% cell-size reduction compared with traditional 6- or 8-transistor memory technology, two orders of magnitude lower standby power, and up to 75% lower operating power. It can also be manufactured using mainstream integrated circuit fabrication processes, significantly lowering manufacturing costs and barriers to adoption. On-chip memory density has continuously increased at a rapid pace. The area occupied by memory, which is currently dominated by 6- or 8-transistor static random access memory is already greater than that used by logic functions in a typical system-on-a-chip application. There is thus a significant demand for a smaller on-chip memory structure. The startup's memory cell is also suitable for embedded memory applications as it is compatible with existing complementary metal-oxide semiconductor processes. The company's 1-transistor memory cell is bi-stable, analogous to a static random access memory cell. This is in contrast to a dynamic random access memory cell which only has a single stable state, and thus requires a periodic refresh operation. Therefore, both states in this company's memory cell are stable while power is applied to it. Currently, the startup has more than 40 U.S. and international patent applications, of which 23 have been granted.

Deterioration of patent notice, industry-specific factors, rewards for litigation, a general increase in litigiousness, the rise of patent "trolls", and the declining quality of patent examination are driving this surge in the cost of litigation. U.S. companies lose $200-250 billion annually due to worldwide copyright, trademark, patents, and trade secret

---

[125]Heller, Michael - The Gridlock Economy: How Too Much Ownership Wrecks Markets, Stops Innovation, and Costs Lives, 2008

infringement[126]. In 2011, litigation by patent trolls cost defendants $29 billion. The patent system, intended to foster and protect innovation, is generating waste and uncertainty that hinders and threatens the innovative process due to the inherent tension between rewarding some innovators while potentially inhibiting the activities of others[127].

Granting patent rights to vital knowledge assets will stifle innovation and foster monopolies that hoard scientific and medical discoveries to the detriment of society. A country's relative productivity depends on its ability to absorb from a common pool of knowledge into developing its domestic technology. The primary spur to technology diffusion and economic growth is not capital resources but the strength of a country's IP protection along with its education and infrastructure. Without strong IP rights, leading countries tend to have insufficient incentive to invent and follower countries tend to have an excessive incentive to copy. Few companies outside the pharmaceutical, biotech, and chemical technology sectors use patents for other than defense, and see patents as strategic assets. In electronics a patent rarely covers an entire product, in a health technology company it often does. A chip may incorporate technologies from dozens of patents, if litigated the industry screeches to a halt. Broad cross-licensing agreements saved the day; cross-licensing is important with respect to technology standards. For small design firms and non-fab firms it is crucial to have patents in order to secure investment, to allow product innovation, and to be acquired[128].

The first-to-file versus first-to-invent debate was finally settled by the new patent reform law. Determining who was first to conceptualize an idea is often difficult but it is easy to determine filing date. If small inventors take longer to prepare patent applications, they might lose out to better financed rivals in this first-to-file world. The earliest days of an industry are critical in affecting how an industry evolves, and firms that can get a head-start in an emerging industry are in the best position most of the time to survive an industry shakeout that inevitably follows. When patents are awarded to the wrong firms during the critical early days of an industry, distortions brought about by such miscues may be long-lived.

The U.S. Patent and Trademark Office (USPTO) defines its primary job in terms of reducing the time it takes to process patents, and continues to reward examiners based on productivity. This encourages granting patents rather than granting only valid patents. If

---

[126]Choate, Pat - Hot Property: The Stealing of Ideas in an Age of Globalization, 2007

[127]Jaffe, Adam and Lerner, Josh - Innovation and Its Discontents: How Our Broken Patent System is Endangering Innovation and Progress, and What to Do About It, 2004

[128]Rivette, Kevin and Kline, David - Rembrandts in the Attic: Unlocking the Hidden Value of Patents, 1999

the USPTO is serious about patent quality, it must overhaul its compensation structure so examiners are rewarded for denying applications on non-novel inventions and for making those denials stick. In the software arena, it should grant patents only on novel, non-obvious software and should require the applicant to describe the covered software in some detail so patents go to only people who have created something not to those who merely thought about creating something. Some argue that patents are fundamentally inapplicable to software, that software development is so cumulative that it is impossible to sufficiently parse out the contribution of one developer, to grant patent rights. It is also counterproductive to do so because subsequent development will be hampered. These ideas are embodied in the open source movement which opposes software patents.

In biotech, the patenting of research tools creates the fear that researchers would need to pay royalties on multiple distinct research tools to market a given product which is retarding or will soon retard the inventive process. If patents are granted only on novel, non-obvious inventions then researchers will have to pay royalties to others only for the use of research tools that were truly invented by a patent owner. The relative inexperience of the USPTO with these new technologies combined with their critical importance to the economy has made broader more fundamental problems with the system most evident. Ambiguity and complexity that attends the current process taxes the innovation process, and small companies and startups may suffer disproportionately, resulting in fewer experiments with novel combinations of knowledge in our society.

Early Adopters

The ultimate purpose of a business, of economic activity, is to create a customer. Some argue that the West grew rich first because people here are more open to new technology, and possess a greater absorptive capacity for innovations. Entrepreneurs often launch their innovations first and discover the benefits and costs afterward[129]. Both developers and the users of innovations require a high degree of entrepreneurial resourcefulness in problem solving. Consumers in America seem more willing to splurge on new things, and the public seems more inclined to assess the level of risk in terms of not only potential negative consequences of technology but simultaneously in terms of potential benefits. Moreover, the users of products and services are increasingly able to innovate for themselves. It is becoming easier for them to get precisely what they want by designing it for themselves. As previously indicated in Chapter 2, users were the developers of about eighty percent of the most important scientific instrument innovations, and developers of most of the major

---

[129]Phelps, Edmund – Economic Justice and the Spirit of Innovation, 2009

innovations in semiconductor processing as were most widely licensed chemical production processes[130].

Natural, Renewable, Biodegradable
Materials from Agricultural Byproducts and Mushrooms

The early adopters here are possibly the U.S. automotive and structural core industries. This startup based in Troy, New York is being provided federal funding from April 2012 to March 2016 in the amount of approximately $1.0 million for R&D work to quantify the mechanical performance of bio-composites sourced from fungi while concurrently scaling and demonstrating material production. The engineered- composites market continues to grow steadily because of the high strength-to-weight and stiffness-to-weight ratios of these systems, as compared to conventional engineering materials. Engineered woods are ubiquitous in the construction and furniture industries, but due to domestic indoor air quality regulations, these materials are being phased out or these industries are forced to use expensive formaldehyde-free adhesives. Similarly, the automotive industry is under regulatory pressure to find alternatives to fire-retardant foams that cannot be recycled due to inorganic filling agents. The technical results have so far demonstrated bio-composite materials that can compete both economically, and on mechanical performance, with the aforementioned competitors, while meeting legislative demands. The startup is working with key industry partners to meet performance metrics and demonstrate quality pilot production.

The company harnesses the power of nature, the cleanest technology on Earth, eliminating the pollution generated across the petroleum-based plastics supply-chain. Unlike plastics, which come from unsustainable petrochemicals, mushroom materials start with plant-based farm waste and can end up in the garden, fully compostable. The startup's award-winning, patented materials technology provides a platform of possibilities, blending feedstock and adjusting biological production parameters to meet cross-industry performance specifications. This effort, if successful, will result in a customizable bio-composite for a broad range of markets, including automotive, transportation, architectural, furniture, sports, and recreation. These materials are truly sustainable, since both the laminates and cores used in the sandwich structure consist of renewable materials. They also require significantly less energy to make than other biocompatible composites, because the material is grown instead of synthesized, and the material is completely compostable at the end of life. Furthermore, this startup will demonstrate and scale the best manufacturing processes to a pilot stage capable of manufacturing high volumes of quality product. Since these materials leverage regional lignocellulosic byproducts from domestic agriculture and industry, a regional manufacturing model is presently being pursued to reduce transportation and feedstock costs. This will not only bring additional value to U.S. agricultural markets, but will spur rural economic development through domestic manufacturing. Finally, these advanced biological materials represent a new paradigm in manufacturing, offering safe, biodegradable alternatives to traditional petroleum-based ones.

The willingness and ability of society to effectively use products and technologies derived from scientific research is far more important than owning a share of such research. New firms start from scratch and thus have to develop knowhow and products, customer

---

[130]Hippel, Eric – Democratizing Innovation, 2006

relationships and organizational structures. They have only so much time and money to accomplish the mission. This has significant influence on how such businesses innovate[131]. Early adopters interacting with innovators allow the U.S. economy to benefit from high-level scientific research and technology wherever developed. It is often better to protect the interests of such consumers of innovations who are not too hard to sell to. Once you have a product that is competitive and have your first adopters there is a need to go from a tech-centric to a market-centric focus. The role played by consumers in the multiplayer innovation game has been frequently neglected. Users play an entrepreneurial role in the design of new products, bearing unmeasurable and unquantifiable risks and developing ground-level knowledge. Venturesome users play a critical role in determining the ultimate value of innovations. Such customers also encourage innovators to optimize offerings for their needs and to invest in marketing to them. For example, information technology users in the U.S. are exceptionally venturesome - deployment of innovations by such early adopters made a significant contribution to the nation's continued prosperity. American businesses are stingy in conventional investments but exceptionally daring when it comes to information technology. This eagerness of Americans to buy and use information technology encourages innovators to develop products optimized for the U.S. and to devote significant resources in marketing to U.S. customers.

Technology-specific Networks

To explain the need for technology-specific networks we consider an example SBIR-funded startup featured in this subsection whose initial focus was to raise sufficient capital. The founder's earlier experience with the acquisition of his previous startup had already created for him a powerful network of technology-specific investors and potential customers. It soon became clear that venture capital firms were not keen on investing in the electronic design automation (EDA) space. However, the company's pedigree and technical competence indicated there was likely more value to the technology than what was being recognized by the venture market. Feedback from potential customers, and obtaining a purchase order was simultaneously sought to assist not only with revenue generation but also for market validation so that potential investors would then look more favorably upon the firm.

A leading EDA vendor was engaged to evaluate the toolset developed by this company. The vendor appreciated its technical elegance, however more importantly, the interaction indicated that the silicon-IP (intellectual property) market, with consistent growth rates,

---

[131]Bhide, Amar - The Venturesome Economy: How Innovation Sustains Prosperity in a More Connected World, 2008

might provide the greater economic opportunity compared to the relatively flat-growth EDA industry. Furthermore silicon-IP companies

---

Electronic Design Automation
Simulate Synchronizer Behavior in System-on-Chip Design

This publicly-funded research and development work from September 2009 to August 2014 in the amount of approximately $1.1 million to a startup based in St. Louis, Missouri was to develop and apply a novel design methodology to confront problems associated with deep sub-micron, system-on-chip integrated-circuit (IC) designs. This R&D effort developed design services for companies wishing to market complex, proprietary, low-power integrated circuits using a unique design tool, one that applies a mathematically sound approach to the production of large, hazard-free, network-on-chip products. The goal for this tool is to reduce traditional design cycles by eliminating most of the global verification effort while improving design robustness. This R&D effort will reduce design costs, time-to-market, and power consumption in semiconductor devices. This suite of tools can significantly increase the productivity of integrated-circuit design engineers, reduce power consumption of electronic control, communication and computational systems, and increase U.S. competitiveness against off-shore system-on-chip designers, particularly with respect to low-volume products. This endeavor is important to the future of the national electronics marketplace - without a major reduction in the time spent on global verification, the benefits of higher levels of integration, including reductions in time-to-market, conservation of power and increases in reliability, will not be available to many important electronic market sectors. The IC designer's challenge is to simultaneously deliver increased processing complexity, greater speed and manage power consumption reliably and consistently. The answer lies in optimizing the communications between the integrated circuit's component parts. Within each IC design there are defined communication pathways between the components including point-to-point, one-to-many, and many-to-one connections, some unidirectional and some bidirectional. In current designs, these communication paths are synchronized by a global clock tree, a structure which has extremely stringent timing requirements and is increasingly difficult to manage. The clock tree also takes up increasing amounts of space and consumes more and more power. As communication speed increases and the paths between components get longer, it becomes physically impossible to meet the stringent timing requirements imposed by the clock without adding timing buffers. However, when these are added a number of unwanted effects are introduced such as increased latency, decreased performance, increased power consumption, and increased overall design complexity. Additionally, many components operate at different speeds, thus forcing a split into multiple clock domains, each requiring one or more clock domain crossings. The heart of a clock domain crossing is a synchronizer, used to make sure data accurately flows from one clock domain to another. To satisfy power constraints in these synchronizers the distinction between a 0 and a 1 has become so small it is increasingly difficult to reliably determine which state a synchronizer is in. This makes it more difficult for synchronizers to perform their function resulting in increasing risk of failure due to metastability and the problem of timing closure to maintain reliable communications between components.

---

have had venture backing and have been acquisition targets. A leader in this market indicated interest but preferred to wait to see more progress on the technology development front. The startup therefore focused their existing technology efforts towards the automated creation of IP cores that could be used in the silicon-IP industry. The technology was tested on a field programmable gate array (FPGA)-based product and provided the best quantification of the technology's value proposition to date. These FPGA performance increases indicated that there would likely be similar improvements realized in the higher-volume application specific integrated circuit market.

This startup's success will in part be attributable to the company's early engagement with customers as well as a willingness to adapt their go-to-market strategy. While the company sought to approach the market with two EDA tools, providing a modest opportunity, it was their interaction with the market, its access to networks specific to chip design and its entrepreneurial thinking that has positioned it for a much larger opportunity, the silicon-IP market. In 2011, the company attended a semiconductor forum at Stanford University. Representatives of industry, venture, and academe specific to the semiconductor industry were in attendance. Additionally, Stanford's location in the heart of Silicon Valley provided an opportunity to hold a market focus group for this company with key industry players to pose questions and express the needs of the semiconductor industry. This opportunity allowed the startup to refine its go-to-market strategy as well as providing additional insights into semiconductor technology pain-points that might be successfully addressed by other such taxpayer-funded small businesses.

Industry

Companies reside on the leading edge of the drive for successful innovation, the prime venue where new knowledge is converted into useful products, and where success and failure can be gauged in terms of patents, market share, sales, profits, stock prices and the like[132]. The old approach to innovation was based on a social bargain with large companies - give us a monopoly in our markets and we will invest in basic R&D. That old industrial lab model followed by large vertically-integrated companies such as Xerox and Bell Labs is essentially defunct[133]. Companies like Intel, Oracle, Cisco, Genentech, Amgen, and Genzyme conduct little or no basic research on their own, and innovate with the research discoveries of others. The general principle in leveraging external technology is to utilize internal and external ideas to create value for your customers and to rely on internal technology and assets to claim a portion of that value. Proctor & Gamble, for example, now sources 50% of its innovations from outside the company.

---

[132]Buderi, Robert - Engines of Tomorrow: How the World's Best Companies are Using Their Research Labs to Win the Future, 2000

[133]Gertner, Jon - The Idea Factory: Bell Labs and the Great Age of American Innovation, 2012

Open innovation implies valuable ideas can come from inside or outside the company and can go to market from inside or outside the firm as well. It expands the role of internal researchers to include not just knowledge generation, but also knowledge brokering. Companies become active buyers and sellers of IP, and instead of managing their IP to exclude rivals, they manage it to profit from others' use of it. It is fairly typical from the perspective of most IP sellers to believe that they have done all this work for years now, and they would like to obtain some return on the investment they have made in the intellectual property that they have created. Similarly it is fairly typical from the perspective of most IP buyers to ask themselves the question - how much more would they have to invest in the IP they plan to acquire to derive commercial value from it[134].

The approaches to managing innovation by industry are exemplified by the following three companies: IBM retreated from atoms and molecules and advanced toward software and solutions, and today makes extensive use of others' technology in its business. It creates programs to bring external technology inside and implements an aggressive program of investing corporate venture capital in startups to extend its markets. Instead of reinventing the wheel and abiding by the not-invented-here syndrome, IBM uses open innovation to build new solutions and services for its customers and makes money doing it. IBM has a powerful portfolio of semiconductor patents, and its network of agreements and strong internal IP makes it a safe foundry for new firms seeking to enter this industry. Ownership of extensive IP gives the company an edge in competition to supply complex products where the possibility of IP infringement is real and hard to discern in advance. Intel leverages external technology through careful monitoring of academic research and through corporate venture capital investments in startups developing promising new technologies. Lucent embodies a third approach to managing innovation – taking internal knowledge out to the external market. Lucent's New Ventures Group, a halfway house to enable startups not ready to obtain outside venture capital to develop their ideas further within Lucent.

Open innovation companies accept that the myriad startup firms funded by venture capital will be an enduring part of the innovation landscape. Startups thus financed are pilot fish for potential market opportunities, because these new firms are selling real products to real customers. Industry can exploit venture capital's ability to fund multiple organizational experiments to commercialize technologies, giving it tremendous visibility into the business models of new and emerging companies. The knowledge of business and markets may give valuable insights into unmet needs that a large company cannot or chooses not to address internally. Unfettered

---

[134]Chesbrough, Henry - Open Innovation: The New Imperative for Creating and Profiting from Technology, 2003

research is no longer a logical or necessary investment for a large company for it can obtain the fruits of this research by the acquisition of appropriate startups. Now, internal production lines and deft, aggressive engineering rather than in-house scientific breakthroughs is what is required. The models of innovation in industry are changing. Besides funding external research and licensing, there is venture investing and technology acquisition, each evaluated in terms of timing, its risks, and its rewards.

Revealing Nanoscale-level Properties
Infrared Chemical Nanospectroscopy

This federally-funded research and development work from April 2010 to September 2015 in the amount of approximately $1.0 million to a startup based in Santa Barbara, California was to develop infrared nanospectroscopy, leading to the first commercial instrument capable of dramatically improving the resolution and sensitivity of chemical imaging at the sub-20 nanometer (nm) scale. Conventional infrared spectroscopy is the most widely used technique for chemical characterization, but spatial resolution limits have prevented it from being applied at the nanoscale. The atomic force microscope (AFM) has excellent spatial resolution, but until recently had no ability to perform chemical spectroscopy. Infrared (IR) AFM has demonstrated spectroscopy at well below conventional diffraction limits although the current spatial resolution and sensitivity are on the order of 100-200 nm, and the method requires specialized sample preparation. This R&D effort combined simulations with the development of experimental techniques and prototype instrumentation to enable commercialization of infrared spectroscopy and chemical imaging down to the scale of single monolayers and individual molecules. The impact of this work will be to provide researchers a robust capability to leverage the power of infrared spectroscopy over broad wavelength ranges and at resolution scales well below current limits. With billions of dollars of global investments in nanoscience and nanotechnology, the lack of IR nanospectroscopy technology leaves an enormous gap in needed characterization capabilities. This novel AFM-IR platform will enable a wide range of high-resolution characterization methodologies in materials science and life sciences including correlation of morphological, chemical, mechanical and optical properties. The tools being developed by this startup reveal hidden chemical and mechanical characteristics of materials through spectroscopy and thermal analysis at nanometer scales. With a researcher's productivity in mind, the company delivers integrated hardware and software solutions that clear paths to their next discoveries. Together, infrared spectroscopy, and thermal and mechanical analyses provide a special dimension to AFM imaging. The startup recently introduced their second generation nanoscale IR spectroscopy platform. Featuring top-side illumination, this tool greatly expands the range of samples that can be studied. It combines the nanoscale spatial resolution capabilities of AFM with infrared spectroscopy's ability to characterize and identify chemical species. Understanding structure-property correlation, especially for samples with spatially varying physical and chemical properties, is critical in a diverse range of fields, including polymers, materials science, life sciences, semiconductors, and data storage.

The above featured startup illustrates the case of the founding team possessing a deep understanding of an industry that they had worked in for over a decade. Contemplated issues included the belief that this fledgling firm would first need to launch an accessory device for the smaller atomic force microscopy (AFM) market before it could complete its standalone device. Soon thereafter it was determined that this startup could launch both the instrument and the accessory provided it partnered with the leader in the AFM market. Was there any downside to partnering with this market leader for the accessory device which would be an addition to that company's AFM machines while launching its own standalone machine for the infrared (IR) instruments market? Another key issue was the need for a strategic investment with the right-of-first-refusal. A corresponding term-sheet was required from this market leader.

The management team engaged two experienced term-sheet negotiators, one with a corporate venture capital background and the other with experience founding multiple startups. The immediate issue facing the small business was whether it should partner with the larger company in launching its product or go it alone. Their technology breakthrough is in the area of nanoscale IR spectroscopy which positioned the startup between two instrumentation sectors, the $1B/year IR spectroscopy sector and the $250M/year AFM sector. The two mentor negotiators spent time with the startup to sort through the pros and cons of a partnership, its timing, and its implications on company valuation. These discussions were crucial in clarifying issues. The company founders, with their in-depth knowledge of this industry sub-segment, eventually decided to go it alone and launched their product, the nanoIR. This company has not precluded a partnership with market leaders - if it did end up going that route, it would be in a much stronger position to negotiate terms in its best interest.

As stated earlier in Chapter 2, products based on disruptive technologies are typically cheaper, simpler, smaller, frequently more convenient to use, generally promise lower margins not greater profits, and are first commercialized in emerging or insignificant markets. Leading firms' most profitable customers generally do not want and indeed initially cannot use products based on disruptive technology[135]. The essence of strategic technology management is to identify when the point of inflection on the present technology's S-curve has been passed, and to identify and develop whatever successor technology rising from below will eventually supplant the present approach. It is this upward mobility that makes disruptive technology so dangerous to established firms and so attractive to startups. The promise of up-market margins, simultaneous up-market movement of many of the company's customers and the difficulty of cutting costs to move down-market profitably together create powerful barriers to downward mobility. Firms that attempted to develop radically new technology almost always tried to maintain simultaneous commitment to the old, and almost always failed[136]. Disruptive technologies facilitate

[135]Christensen, Clayton - The Innovator's Dilemma: When New Technologies Cause Great Firms to Fail, 2005

[136]Utterback, James - Mastering the Dynamics of Innovation, 1994

emergence of new markets - there being no one billion emerging markets, it is precisely when emerging markets are small when they are least attractive to large companies in search of big chunks of new revenue that entry into them is so critical. Small markets cannot satisfy the near-term growth requirements of big firms. One way to match size of organization to size of opportunity is to acquire a rapidly growing startup within which to incubate the disruptive technology. The key characteristic of disruptive technology is that it heralds a change in the basis of competition. Attributes that make disruptive products worthless in mainstream markets become their strongest selling points in emerging markets. Because disruption rarely makes sense during the years when investing in them is most important, conventional managerial wisdom at established firms constitutes an entry and mobility barrier that entrepreneurs and investors can bank on.

Manufacturing Partners

Discovery and key inventions across a number of market sectors such as energy storage, power generation, and robotics have taken place within the U.S. innovation system, but manufacturing has moved offshore[137]. Early and substantial investment in process technologies and the actual scaling up of optimized production capacity is essential to attaining large market shares in the increasingly technology-based global economy. Once a new technology is initially commercialized, simultaneous scale-up of production capacity and product differentiation for multiple markets become critical issues. For the U.S., scale-up has become a significant barrier to long-term economic growth. This is because the vast majority of the economic benefits from new technologies results from the growth of their markets after they have been first introduced. Today scaling is becoming much more complex.

The rule of thumb most manufacturers follow is that for every dollar spent on researching a new technology, $10 will have to be spent on developing the product; and for every $10 spent on product development, $100 will have to be invested in the manufacturing and marketing capability necessary to introduce the product to customers[138]. Some of our startups produce offshore because they find that the U.S. simply lacks the supply-chain capacity, technical skills, and the right investment climate for even pilot-scale let alone high-volume manufacturing. How to obtain appropriate domestic supply-chain information and how to connect to global distribution networks in a time-efficient manner are not obvious. As a result, America is finding

---

[137]Tassey, Gregory - The Disaggregated Technology Production Function: A New Model of Corporate and University Research, 2005

[138]Smith, Douglas and Alexander, Robert – Fumbling the Future: How Xerox Invented, Then Ignored the First Personal Computer, 1999

it increasingly difficult to capture economic value generated by its tremendous public and private investments in R&D.

A significant barrier for high-tech small businesses is simply getting on the radar of potential customers that they can later build into a financial engagement. Even if such an opportunity were to become available, the startup often lacks the resources to support the costs involved in the technology development, customization, and demonstration activities required by the potential strategic partner. Even if plugging into such open innovation activities is often critical for the startup, it is hard to convince large companies to financially support untested, new technology. Manufacturing companies want proven "drop-in" solutions, and want startups to bear the cost of most technical risks, large or small. They also want exclusivity on the use of the technology at terms detrimental to the startup's ability to seek other opportunities if ongoing efforts lead to negative outcomes.

Manufacturers are, if at all, only willing to invest small amounts leading to considerable time spent in fundraising. Additionally, contracting terms are such as to leave a startup a considerably small portion of the value created. There is also a lack of innovation in providing access to resources for pilot-scale manufacturing. Venture capital is difficult to attract for it is hard to compete with software startups, and other sustainable sources are currently not apparent. Additionally, the founders' vision of building a manufacturing company does not align with the investors' desire for the startup to grow rapidly and exit with large gains. Overseas sources are increasingly financing such U.S. startups and in the process increasing their manufacturing competitiveness at U.S. taxpayers' expense.

The critical role of startups in creating new and advanced manufacturing technologies is not sufficiently recognized. Current efforts appear to be directed at traditional manufacturing industries and may well help them retain their present global market position. It is uncertain though that these will create a significant advanced manufacturing base any time soon. By far, the strongest resource to nurture the growth of early-stage advanced manufacturing companies is to increase the synergy between these firms and established U.S. manufacturers. This kind of startup/ established firm interaction is common among leading information technology companies and in the life sciences but is less common in the manufacturing sector. These interactions often involve activities that are not directly related to financing and are frequently vital pre-conditions to early-stage and pilot-plant development. These activities include concept validation that draws upon the broad market and production operations of leading manufacturers, use of specialized facilities, business and production mentoring, and early beta customer partnerships. These interactions can lead to investment relationships from venture operations of major companies. Policies should explore the tax treatment of investments by leading manufacturing firms in startups. It is not clear whether current

R&D tax credits actually support the transition from R&D to manufacturing. In general, there is thus a lack of government programs that incentivize or assist small businesses to transition from R&D to manufacturing. Perhaps the solution lies not at the federal but at the state level.

More Efficient Manufacturing
Minimum Quantity Lubrication for Metal Forming Applications

This publicly-funded research and development work from January 2011 to June 2014 in the amount of approximately $734,000 to a startup based in Detroit, Michigan was to develop next-generation supercritical carbon dioxide (CO_2) metalworking fluid (MWF) technology for highly demanding metal forming applications. The approach is to deliver specialized environmentally-friendly lubricants with supercritical CO_2, achieving tool wear rates, forces, and surface finish at least as good as aqueous-based MWFs that are currently in use. It is anticipated that a much smaller amount of MWFs will be required with this technology. This R&D effort involves the formulation of new supercritical MWFs and the optimization of flowrates of oil and CO_2 for metalworking processes. The patented supercritical CO_2 system, the so-called CHiP Lube, is evaluated in actual industrial settings to confirm its capability to replace current MWFs. The effectiveness and efficacy of the CHiP Lube system will also be scaled and applied to other common industrial metal working processes such as rolling, extruding, and cutting.

This R&D effort has the potential to provide an environmentally-benign lubricant system as an alternative to conventional MWFs with equal or better performance and lower cost. At any given time, approximately two billion gallons of MWFs are in use in America. This represents a massive waste stream that must be treated and remediated. In addition, the negative effects of MWFs on worker health and safety are well documented. The components of CHiP Lube are naturally occurring and used in extremely low quantities. Therefore, the waste treatment and worker health concerns are minimized. CHiP Lube has been demonstrated in simple metal removal applications as providing lower tool wear and/or higher machining speeds than conventional MWFs, thereby leading to a lower overall cost of manufacturing. In addition, no carbon dioxide will be produced to run the process, as the CO_2 used will be recovered from other industrial processes such as ammonia and ethanol production. Metal- working fluids used today compromise cooling for lubrication - add oil and you reduce cooling; add water and you reduce lubricity. Oil-in-air minimum quantity lubrication can lubricate but does not provide efficient cooling; liquid nitrogen can cool but does not lubricate well. This firm's supercritical carbon dioxide gives maximum cooling and lubrication potential at the same time, increasing productivity and reducing system-level costs. The startup's technology provides an advanced heat management system that offers benefits not achieved by any other metal-cutting technology available today.

Experience gained and lessons learned from the portfolio of startups with manufacturing needs have shed light on four available pathways:

- No choice but to manufacture overseas
- Acquired by a foreign multinational that in turn absorbs the new technology into their own manufacturing lines

- Acquired by a U.S.-based manufacturer, an increasingly rare outcome
- Help the startup partner with a large U.S. corporation to set up a pilot-plant using the partner's under-utilized manufacturing facilities at subsidized rates

The U.S. has numerous applied-research programs but lacks a systematic institutional focus on developing manufacturing industries at scale for new technology products. It is essential that we expand the resources available for early-stage growth and accelerate startup interaction with major manufacturers to create a more robust environment for the production of products derived from advanced technologies developed by U.S.-based startups. Opportunities as those offered by under-utilized manufacturing facilities should be expanded nationwide to support startups emerging from federal advanced manufacturing research programs. This support could include expanded mentoring initiatives, efforts to mobilize corporate partners interested in supporting beta testing of products, and/or the creation of a seed fund for advanced manufacturing startups. A focus on creating a national network to foster stronger startup interactions with leading manufacturers would be a key element. This network would complement and aid the efforts of individual universities to integrate a manufacturing focus into innovation ecosystem programs to create and support startups[139].

Investors

Amazon, Yahoo!, Cisco, Intel, Microsoft, Genentech – in all these cases venture capitalists provided funding, contacts, reputation, and advice while these companies were in their formative stages. Total venture capital investment accounts for less than 0.15% of gross domestic product (GDP), and less than one percent of new businesses started each year in the U.S. receive venture funding. Venture-backed companies create eleven percent of all private sector jobs. They generate annual revenues equivalent to twenty-one percent of GDP. The dozen largest technology companies were all venture-backed. Together those twelve companies are worth more than $2 trillion, more than all other technology companies combined[140].

Table 5.1 indicates the amount of capital managed by venture capital firms in the United States over the last five decades. Most high-tech entrepreneurial ventures have a number of fundamental problems that make them difficult to finance through some combination of

---

[139]President's Council of Advisors on Science and Technology: Report to the President - Capturing a Domestic Competitive Advantage in Advanced Manufacturing, 2012

[140]Thiel, Peter and Masters, Blake – Zero to One: Notes on Startups, or How to Build the Future, 2014

angels, venture capital, private equity firms, foundations, endowments, corporate venture units, and crowdfunding platforms[141]:

- Uncertainty about the future
- Information gap between inventor/entrepreneur and investor, and how to value the invention or what it would take to get to market
- Soft assets – trade secrets, tacit knowledge, copyrights, and patents
- Uncertainty around market conditions

Table 5.1   U.S. Venture Capital Industry

Year	# of VC Firms	Capital Managed ($B)
1970	28	1
1980	87	4
1990	100	28
2000	1053	224
2012	522	199

Besides reliable financial support, inventors and entrepreneurs need talented management, and coaching insight from professionals who specialize in innovation and taking products and services to the market. Table 5.2 provides a list of attributes that venture capital firms and startups together typically consider, after the startup team has been fully vetted, before making/accepting an investment. Many venture capitalists conceive of their funding decisions as investments in business models – they give active consideration to a variety of possible business models and work with their portfolio companies to adopt one that seems to fit well in a particular venture. They provide careful governance and oversight to select a more promising model, rejecting models that no longer seem to be effective. Most entrepreneurs do not see clearly the risks facing their business and many have doomed their fledgling company accepting money from the wrong investors. Entrepreneurs often feel that the terms demanded by venture capital firms are far too onerous for the amount of capital provided, especially when they demand strict control rights and large equity stakes.

The venture capital model in the U.S. is broken; what is missing is the right kind of investing – smart, patient, and long-term. As can be inferred from Table 5.3 which provides a list of the top fifteen venture capital firms by assets ($M), the size of their early-stage

---

[141]Gompers, Paul and Lerner, Josh - The Money of Invention: How Venture Capital Creates New Wealth, 2001

investments ($M), early-stage deal count and the industries they invest in, that we have a possible case of market failure in that venture capitalists (VCs) are not funding strategic industries – sectors that for example gave rise to DEC, Intel, Genentech, and Amgen. More venture investment, approximately $11 billion, flowed into software companies in 2013 than at any point since 2000, which was nearly the

Table 5.2   **General Investing Considerations**

Business Model	• Product versus service versus process • Licensing
Growth Model	• Lifestyle (no growth) • Bootstrap (slow growth) • Strategic partnering (medium growth) • Private equity (high growth)
Financing Model	• Timing and amount (best to raise capital after critical milestones achieved) • Valuation (tangibles, intangibles (IP)) • Dilution/number of rounds • Negotiating terms • Legal advice • Use of brokers
Investment Sources	• Public – equity-free, non-dilutive • Friends and family • Institutional - VCs/angels/private equity; corporate venture capital; university endowments; foundations
Technology	• First valley of death • Second valley of death – pilot-scale; scale-up (high-volume manufacturing)
Exit Strategy	• IPO; reverse IPO • M&A • Private equity sale

peak of the last technology bubble. Software startups are eating up 37% of all VC funding, the highest percentage since 1995 when such data on this subject was gathered by PricewaterhouseCoopers. The phenomena of VC-"herding" is seen in the desire to pour money into social networking, e- commerce, and online-games while investments in semiconductors for example is close to a twelve-year low. Investment dollars must be steered to the right places – the recent $19 billion acquisition by Facebook of WhatsApp can be seen as a corrosive counter-example.

Venture funding is now concentrated in a few industries such as information technology, healthcare, and internet-related companies. Many promising firms in other industries are thus not attracting venture capital. VCs are no longer concerned with helping to build great companies and new industries. They are simply not prepared to invest

Table 5.3   Early-stage Investments by VC Firms

Name	Location	2014 Assets; investments; deal count	Industries
Andreessen Horowitz	Menlo Park, CA	4,350; 1,020; 50	Consumer products/services; software
Khosla Ventures	Menlo Park, CA	3,100; 809; 45	Software
SVAngel	Palo Alto, CA	1,040; 736; 47	Commercial services; software
Accel Partners	Palo Alto, CA	9,600;722; 29	Communication/networking; durables; healthcare systems; media; software
New Enterprise Associates	Menlo Park, CA	13,000; 691; 44	Communication/networking; energy services; healthcare devices & supplies; IT services; software; pharma/bio
Sequoia Capital	Menlo Park, CA	10,000; 650; 30	Commercial services; restaurants; communication/networking; hotels/leisure; semiconductors; bio; pharma; software
Venrock	Palo Alto, CA	2,600; 620; 15	Communication/networking; pharma/bio; software
First Round Capital	San Francisco, CA	633; 606; 34	Commercial services; software
Spark Capital	Boston, MA	1,825; 542; 18	Media; software; transportation
Triangle Peak Partners	Carmel, CA	516; 535; 3	Software
Franklin Square Capital	Philadelphia, PA	1,100; 505; 1	Apparel and accessories; commercial products; energy services; utilities
Kliener Perkins	Menlo Park, CA	6,833; 490; 33	Commercial services; computer hardware; medical devices; software; financial services; pharma/biotech
Google Ventures	Mountain View, CA	1,200; 462; 35	Software
Founders Fund	San Francisco, CA	2,369; 450; 21	Commercial services; media; financial services; software; pharma/bio
ARCH Ventures	Chicago, IL	1,500; 446; 9	Commercial services; healthcare devices and supplies; pharma/bio

in industry-changing opportunities, and we are thus at a tipping point of losing dominance in many industries all at once. With their short-horizons and numbers-only approach, flipping startups for a quick buck, VCs are becoming mere extractors of profits than builders of lasting businesses that benefit society. Many VC firms say they would not fund a company building hardware or developing hard science or involving capital-intensive plays in cleantech or any of the tough physical sciences because it is just too hard and expensive. Semiconductors, computers, energy, telecommunications, materials, robotics – VCs do not typically do these anymore[142]. Startups are demanding more assistance from their VCs and these are getting more complex. For instance, in many industries strategic alliances with large corporations are now essential to startup success, but these agreements entail complicated, time-consuming negotiations.

The SBIR source[143] analyzes startup funding from both federal research and development grants and venture capital sources. This helps identify private equity investors that are most actively investing in high-technology firms. Table 5.4 is a list of the top-ten VC firms, ranked

Table 5.4   Top Investors in High-tech

Name	% Portfolio SBIR Supported	Sector (majority of investments)
Harris & Harris Group, New York, NY	50	Defense
Lux Capital, New York, NY	44	Defense
In-Q-Tel, Arlington, VA	30	Science
Syngenta Ventures, Basel, CH	50	Science
MedImmune Ventures	39	Healthcare
CalCEF, San Francisco, CA	50	Defense
Stata Venture Partners, Needham Heights, MA	38	Defense
Osage University Partners, Bala Cynwyd, PA	32	Technology, Life Sciences
Fletcher Spaght Ventures, Boston, MA	40	Technology, Life Sciences
Novartis Venture Funds, Cambridge, MA	32	Life Sciences

according to the percentage of SBIR-backed companies in their historical portfolios and weighted by the number of companies in each portfolio. In the period 1958-1969 the U.S. Small Business Innovation Corporation (SBIC) directed $43 billion to young firms, more than three times the total private venture capital invested. Some of America's most dynamic technology companies such as Apple, Qualcomm, Compaq, and Intel received SBIC and/or SBIR support before they went public. If government programs can identify and support firms in neglected industries they might provide the credibility these high-potential underfunded firms need to succeed. Since the U.S now has an active venture sector, public programs must shift their focus from building innovation infrastructure to two new functions:

---

[142]Is Silicon Valley Investing in the Wrong Stuff?, Wall Street Journal, 2014

[143]https://sbirsource.com

- Assessing a new firm's potential and providing appropriately sized funding

---

Emotion-reading Machines
Cloud-enabled Analysis of Facial Affect

This startup based in Waltham, Massachusetts is being provided federal funding from March 2012 to February 2016 in the amount of approximately $1.1 million for R&D work to commercialize the world's first cloud-based emotion measurement platform. Leveraging this public funding, the company has raised tens of millions of dollars of venture capital. The employee count has increased from 2 in 2010 to 35 in 2015. Today, the majority of market research is expensive and slow, relying either on subjective self-reports or costly, obtrusive lab-based technologies. This emotion measurement platform aims to democratize market research by translating nonverbal facial expressions into intuitive emotional insights. It also drives down research costs and improves market reach through the use of widely available webcams as the means to record faces. This platform enables businesses of any size to capture consumer's emotional reactions as they engage with their brands, particularly in the areas of advertising, product design and packaging. For example, brand managers, marketers and agencies can optimize ad effectiveness by evaluating viewers' tacit, moment-by-moment emotional response, in real-time over the web, and through the platform's emotion norms database. The technical objectives of this R&D work focus on implementing automated facial analysis as a scalable cloud-based software-as-a-service platform, building the emotion norms database, and deploying the platform with leading market research partners. Emotions influence every aspect of our lives - from the way we interact with each other to the decisions we make. Emotions are the number one influencers of attention, perception, memory, human behavior, and decision-making. The human face is a window into our emotions. By reading the face and its many expressions, we interpret emotions and naturally and spontaneously connect and communicate. Deep insight into consumers' unfiltered and unbiased emotional reactions to digital content is the ideal way to judge a contents' likability and its effectiveness. This effort, if successful, will disrupt longstanding methods in market research by objectively measuring people's emotional experiences unobtrusively, in real-time, at scale, and cost-effectively. While this differentiated emotion measurement technology can be leveraged in several target markets, the company's initial focus is on measuring advertising effectiveness and media research to deliver actionable insights to leading media and market research companies. In addition, this cloud-based emotion measurement platform has the potential to significantly accelerate research in behavioral sciences by enabling the crowdsourcing of huge corpuses of naturalistic and spontaneous responses to a wide range of interactions and experiences from online learning to social gaming. It also allows entirely new research questions to be asked and tackled with ecologically valid data, such as whether individuals on the autism spectrum respond differently to content. Thus, in line with the origins of this technology, this startup's product accelerates psychological and clinical research on social-emotional intelligence. The long-term vision for this software-as-a-service platform is to "emotion-enable" the internet, giving consumers and organizations the ability to add emotion context to all online interactions.

- Addressing the spillover problem where other parties benefit more than the new firm which originally created and invested in a new idea, and the original innovator bears all costs associated with R&D but captures only a fraction of the benefits, by subsidizing some of the cost of developing new products or processes by young firms

Even startups not viable as businesses may be attractive investments for the government that wants to encourage innovation across the entire economy. A startup that is developing a new instrument for synthetic biology or genome editing applications, for example, may never create a very profitable business but if this instrument accelerates the work of researchers in industry and academia, the benefits to society may be great. The government can focus on technologies not popular with VCs, or provide follow-on funding to a company already venture capital-funded when the private funding stream is rapidly lessened when scale-up for manufacturing in America is imminently required.

University

American universities are the nation's chief source for the three ingredients essential to continued growth and prosperity: highly trained specialists, expert knowledge, and scientific advances others could transform into valuable new products or life-saving treatments and cures. A high proportion of the leading new industries in America, perhaps as much as eighty percent, are derived from discoveries at U.S. universities. The laser, magnetic resonance imaging, frequency-modulated radio, the algorithm for Google searches, the global positioning system, deoxyribonucleic acid fingerprinting, fetal monitoring, scientific cattle breeding, and tens of thousands of other inventions, devices, medical miracles, and ideas that have transformed the world had their origins in America's research universities[144].

In 1980, Congress passed the Bayh-Dole Act, which made it easier for universities to own and license patents on discoveries made through research paid for with public funds. Today, U.S. academic institutions and national laboratories have become focal points for economic development. The recent wave of entrepreneurial activity by American universities is a response to the reductions in government support for higher education that began in the 1970s[145]. In the case of biotechnology, as the industry boomed, life scientists started to seek patents on their discoveries and began to receive stock from new firms eager for their help and even to found new companies based on their own discoveries[146]. Some believe that active collaboration with colleagues in the biotech industry is actually useful in stimulating basic

---

[144]Cole, Jonathan – The Great American University: Its Rise to Prominence, Its Indispensable National Role, Why It must be Protected, 2009

[145]Bok, Derek – Universities in the Marketplace: The Commercialization of Higher Education, 2003

[146]Kenney, Martin - Biotechnology: The University-Industrial Complex, 1988

---

research. The process of scientific exploration has become a much more collaborative process, requiring input and stimulation from a wide variety of sources. In drug development, working with industry is essential, since pharmaceutical companies often have databases, vast libraries of relevant compounds, sophisticated computer models, and other research materials that university labs do not have and that scientists must be able to use to perform high-quality work[147].

Increasingly, universities will be the locus of fundamental discoveries. Startups require greater access to the results of university research to transfer these discoveries into innovative products and services. It will benefit society if these ideas flow to the market through multiple business models. Restrictions on diffusion of publicly-funded research can be damaging especially since industry is doing less of basic research. We must support broad disclosure of research results, and create win-win agreements between small businesses and universities. Some U.S. universities with a long-term outlook and small upfront fees successfully spinoff startups and derive a large potential upside. The common element in these universities is that they think and act entrepreneurially as an institution. Others license their patents, either exclusively or non-exclusively, to existing small businesses in exchange for downstream royalty payments and/or take a small equity position in these entities. In general universities can be more inviting to startups or small businesses through entrepreneurial incentives such as changes in tenure policies especially for younger faculty, reconsidering, for example, the relative importance of publications and patents. They could more effectively nurture small businesses by improving the ease of access to university technical advisors and experts, by encouraging faculty to take up consulting roles in such companies, and by facilitating the use of university equipment, laboratories, and other research resources.

Venture capitalists often need to be adept at cobbling together startups that buy patent rights from different university labs none of which may have enough intellectual property to go it alone. Their sentinels tend to pick up on important discoveries very early enabling the VC-funded firm to orchestrate patent filings before beans are spilt in scientific journals. The fair return model is where the university asks a potential small business what they believe is a fair return if they successfully commercialize intellectual property. In willing to forego potential short-term licensing revenues an agreed amount is paid instead of a strict upfront license fee. Additionally, universities could grant exclusive licenses for a specific field-of-use rather than a wider field thus opening the technology to a multiplicity of potential products and services that respect that particular field but does not limit broader application.

---

[147]Etzkowitz, Henry - The Triple Helix: University-Industry-Government Innovation in Action, 2008

Investing in startup companies is not a business in which universities have special expertise; state officials ask campuses to

---

Rediscover the Sense of Touch
Force Modulation for Haptic Touchscreens

This publicly-funded research and development work from September 2012 to February 2016 in the amount of approximately $1.0 million by a startup based in Evanston, Illinois is to advance the development of electrostatic surface haptics as a means of providing tactile feedback for touchscreens. In electrostatic surface haptics, a user's fingers are selectively pulled down to a glass surface by an electrostatic potential, just as a capacitor's plates are attracted to one another. The enhanced normal force leads to an enhanced frictional force when the fingers are moved across the surface. By modulating the electrostatic force in response to finger position and time, a great variety of compelling haptic percepts can be leveraged. Low-voltage techniques, high-dielectric coatings, avoiding the necessity of grounding the user, patterns of transparent conductance that allow different fingers to experience different sensations, and software controls were first developed. Then, methods of finger position-sensing that are fully compatible with haptic actuation were developed. The new methods also offer the possibility of discriminating multi-touch fingers according to same or opposite hand, or the fingers of a second person on the same touchscreen. The startup's surface technology offers an enhanced touchscreen experience that lets one feel what one sees on a touchscreen – the edges of keys, the snap of a toggle switch, the swipe of a turned page, and the direction and magnitude of impacts in a game. The small business is the exclusive licensee of a suite of patents developed in one of the world's foremost haptics research laboratories.

The company's founders, intensively involved in the operations of the company, are Northwestern University faculty with a record of successful spin-offs. Haptics is unique among the senses in being bilateral. Consider pushing a toggle switch: the way that the switch feels does not depend solely on the forces your hand applies, nor does it depend solely on the movement of the switch. It depends on the two-way relationship between motion and force. The question naturally arises: can we control the motion-force relationship on a touchscreen or touchpad? The startups' technology provides a set of techniques for controlling the in-plane forces experienced by a fingertip as a programmable function of the finger's motion. The in-plane forces arise from friction, and the company's techniques allow friction to be increased, decreased, even reversed or redirected. The impact stems from enriched communication with touchscreen devices. Already touchscreens are of immense commercial importance. The worldwide market for touchscreen modules grew three-fold from 2009 to 2011 with smartphones, tablet computers, and automotive display applications leading the charge. Consumers have been attracted to the versatility of touchscreens as input devices, yet the absence of haptic feedback has limited accessibility for low-vision populations, exacerbated safety concerns in applications such as driving, and prohibited the aesthetics of touch from being fully developed.

---

speed innovation, job creation, and economic growth by cooperating more closely with industry[148]. In this context, universities tend to think of tech-transfer offices as profit centers, not an optimal role because

---

[148]Stankiewicz, Rikard - Academics and Entrepreneurs: Developing University-Industry Relations, 1986

then these entities often focus on smaller short-term returns and miss the big picture. Tech-transfer is a mechanism developed and enlisted by universities to position themselves three rather than five steps away from the market. Taxpayers do want something more definitive back from their investments than just a larger stock of knowledge. Universities gain political support based on their core mission – teaching, research, and public service – but also for their capacity to be an inventor that offers a return on investment. Another driver is that centers of innovative activity are places where economic development occurs. Since almost all university inventions are derived from non-market-oriented problem solving, the key lies in translation between a university discovery and its market application.

Government

At a time when the developed world is struggling to generate growth, governments and companies alike should acknowledge the fact that they are partners not foes. Years before the great recession, many large corporations cut back their R&D expenditures with some firms closing their laboratories altogether. The consequence is that a growing share of innovation is now occurring in publicly-funded laboratories and in small- and medium-sized firms, many of which are supported with taxpayer-generated research dollars. Over time, the majority of the economic benefits from investment in technologies are realized from scale-up and subsequent attainment of significant global market shares.

The government is good at funding basic scientific research, and at helping fledgling companies, many of which are too small or still too far from profitability, to attract the interest of the venture capital industry. The state provides the legal systems and the basic security that allows startups to operate, educates workers on whom firms depend, and creates the infrastructure that enables companies to get their goods to market. In many industries the public sector is also a key customer, and government bodies have to approve new products before they come to market[149]. It is imperative that government provide regulatory certainty in areas such as renewable energy, environmental and other permitting, intellectual property, and health information technology. It should leverage the local academic, scientific and research base effectively, resist the temptation to over-engineer, recognize the long lead-times involved, and avoid programs that are too small or too big. The earliest phase, the "generic or proof-of-concept" technology research typically occurs a long time before commercialization. Its broad "technology-platform" character provides the potential for multiple market applications. There is significant underinvestment by industry in this early-phase technology research. The lengthy indirect process of realizing economies of scope from new

---

[149] The Economist: Special Report – Companies and the State, February 22, 2014

technologies is no longer competitive in a world economy that conducts over a trillion dollars of R&D per year and is using increasingly efficient mechanisms for managing this investment. Intense technology-based global competition is compressing all technology lifecycles with the result that windows of opportunity are increasingly narrow. Government, with a lower discount rate, the ability to undertake riskier projects, and the resources to support a broad portfolio of long-term research projects must be a major supporter of the elements of complex modern technologies with public-good content.

To capitalize on the globalization of innovation the state should encourage strong interconnections with entrepreneurs and investors overseas, and strengthen and expand research collaboration in biomedicine, energy, environment, pandemics, security, and other shared global challenges. In the great hubs of cutting-edge entrepreneurial activity – Silicon Valley, Singapore, Tel Aviv, Bangalore, Guangdong - the stamp of enlightened proactive government intervention played a key role in their creation, but for each effective government intervention there are dozens even hundreds of failures where substantial public funds bore no fruit[150]. Similarly the venture capital industry in many nations was profoundly shaped by government intervention. Government may have an important role in priming the pump for additional entrepreneurial and venture activity during an industry's inception, and pioneering entrepreneurs and VCs generate positive externalities that benefit others. Signals provided by government awards are particularly valuable in technology-intensive industries. These investments may have a certifying effect by providing a stamp of approval if government programs can identify and support neglected firms that could have important societal benefits.

High-tech innovation is one area where spillovers are commonplace. Research indicates that if a firm earned a ten percent return on its own investment in research, society would be earning fifteen to twenty percent. The difference between social and private returns is especially large among small firms. These organizations are in particular unlikely to effectively defend their intellectual property positions or to extract most of the profits from their discoveries when competing against larger firms. It thus makes sense for government to fund young research-intensive firms even if the direct financial return-on-investment is somewhat less than would be reasonable given the risks absorbed. Government initiatives should directly intervene to boost availability of financing when required, especially in the two valleys of death (or opportunity) between the invention and the proof-of-concept technology development phase and between proof-of-concept and scale-up to manufacture. Government must think beyond

---

[150]Lerner, Josh - Boulevard of Broken Dreams: Why Public Efforts to Boost Entrepreneurship and Venture Capital have Failed – and What to Do About It, 2012

simply handing out money to reducing barriers other than money that entrepreneurs face. We must be open to high-skill immigrants for they

---

Flight Critical Software
Scalable Semantics-Based Verification

The mission of this startup company based in Champaign, Illinois is to use runtime verification-based techniques to improve the safety, reliability, and correctness of software systems. The company has been awarded three grants from NASA for a total of $871,000 to develop software for flight-critical systems that satisfies all Federal Aviation Administration regulations. Runtime verification is a computing system analysis and execution approach based on extracting information from a running system and using it to detect and possibly react to observed behaviors satisfying or violating certain properties. Specifications are typically expressed in trace predicate formalisms, including finite state machines, regular expressions, context-free patterns, linear temporal logic, etc. Runtime verification can be used with regard to these properties to verify the execution of programs, to enforce properties at runtime, to predict property violations not seen in execution, and also to prove programs self-correct on all paths and all inputs using symbolic execution. Since runtime verification analyzes the behavior of programs as they execute, it is often possible to avoid creating complex or too abstract formal models of the system under analysis. This allows for tools that are much easier to use for developers unfamiliar with formal methods, that are lighter-weight than ones based on well-established formal methods, and, notably, that avoid reporting annoying false positives/alarms that static analysis tools suffer from.

Runtime verification is meant as a supplement to traditional unit-based, functional, and integration testing, providing high confidence in the robustness of application behavior traditionally reserved for complex and inaccessible formal methods techniques. Flight-critical systems rely on an ever increasing amount of software - the Boeing 777, for example, contains over 2 million lines of code mostly written in the C programming language. A scalable static formal program verification tool that is able to prove the functional correctness of flight-critical software, limiting any failure of flight critical software to hardware faults, is required. This R&D project seeks to leverage the matching logic verification framework. Matching logic is generic in an operational semantics of a given programming language, so the company also seeks to provide the semantics of a subset of C, called CIL, which is guaranteed to be deterministic. While they already have semantics for the entirety of C, CIL is more representative of flight-critical software, and the simpler, deterministic semantics will result in a more efficient, and thus more scalable, static program verification tool. The firm is also building a new unification-based rewrite engine that will result in a more powerful version of the matching logic framework. In order to make the tool more commercially feasible, the startup will develop new techniques in pattern inference, so that loop invariants and some pre/post conditions can be determined automatically. Finally, the startup will perform a thorough evaluation of its tool on large-scale software with similar characteristics to a flight system.

---

are the source of many new ideas and innovations. Export rules or immigration restrictions on scientists and engineers draw little public attention, but they are an important part of the ecology of innovation. With regard to framework conditions, the U.S. still is the best environment with strong protection of intellectual property rights, temperate bankruptcy laws, well-developed capital markets, and extensive worker mobility, but it is not abreast in tax policy, regulatory costs, and state-of-the-art infrastructure.

Reduce Constraints on Daily Living
Active Wheelchair Driving Aid for Independent Living

This publicly-funded R&D work from April 2012 to September 2015 in the amount of $768,000 to this startup based in Philadelphia, Pennsylvania was to develop an active driving aid that enables semi-autonomous, cooperative navigation of an electric-powered wheelchair both indoors and in dynamic, outdoor environments. It uses intelligent sensing and drive- control systems that work in cooperation with the driver to aid in negotiating changing terrain, avoiding obstacles and collisions. The system allows for higher-level path planning and the autonomous execution of non-linear routes of travel in a safe and efficient manner. The goal is to enable active, safe, and independent living. As an individual begins to lose cognitive, perceptive, or motor function due to age, injury, or disease, this system can augment that loss by interpreting the user's intent and by "seeing" out into the environment on their behalf. This exteroceptive sensing capability is enabled by leveraging the latest in 3D imaging technology. This R&D effort will have a positive effect on the quality of life and independence of the elderly and disabled. Leveraging robotics for personal mobility can help disabled Americans to participate fully in basic endeavors such as employment, education, and other activities of community life often taken for granted. Economically, a serious side effect of the rapidly growing elder population, to over 70 million by 2030, is that it will place unprecedented strains on the U.S. healthcare system. This impact can be moderated in part by enabling individuals to maintain their independence and live at home longer. It is estimated that adding a single month of independence and health to America's elder population would save $5 billion, while decreasing hospitalization and institutionalization 10% would save $50 billion annually. Technology for home-centered approaches to healthcare such as the proposed system is necessary.

In general terms, this startup builds software and systems that interact sensibly with the physical world. The startup was founded as a mobile robotics company. The team has deep experience in developing full-stack autonomous navigation systems for service vehicles. Their navigation stack is a platform comprised of modules for real-time 3D perception, feature detection, map generation, collision avoidance, path planning, and vehicle control. This platform was designed purposefully to enable advanced driver assistance systems for human-in-the-loop cooperative control applications. The company has expertise in real-time 3D computer vision, light detection and ranging, depth sensors, structured light, and other 3D imaging technologies. It has developed perception subsystems for robots and embedded object recognition technologies enabling intelligent products that interact in real-time with their surrounding physical environment. This startup also works with hardware vendors to expand the reach of its sensing systems. Typical technologies the company works with involve combinations of low-level programming on micro-controllers, embedded central processing units and operating systems; hardware accelerated algorithms; cloud services; mobile user-interfaces; and data engineering/analytics.

Support Networks

Support networks play an important role in the creation and growth of technology-intensive startups. These are highly dependent on the geography of innovation clusters in America, where specific combinations of these play a more important role in nurturing the local entrepreneurial ecosystem. Elements of typical startup support networks are presented in Table 5.5.

Table 5.5   Startup Support Networks

Incubators and accelerators
University science parks; innovation centers; and tech-transfer offices
Facilitators: city elders; domain experts; mentors; serial entrepreneurs
Service providers - law firms; payroll/accounting services; consultants; brokers; recruiters; marketing consultants; fundraisers
Market analysts - reports; technology roadmaps; technology-specific blogs
Investors - corporate VC; angels; VC; hedge funds; investment banks; private equity firms; university endowments
Economic development agencies
Innovation portal and networks
Venture forums; small business conferences; technology showcases
Professional societies and conferences; trade groups and trade shows
Local government resources; tax incentives; and state initiatives
Prizes and competitions
Intellectual Property exchanges
Regulatory agencies and standards bodies

The startup featured in the previous page, for example, has had the good fortune to be located in NextFab, a part of the support network in and around Philadelphia set up to assist small hardware firms and those small businesses that have light manufacturing needs. NextFab was founded to help counteract the extensive offshore outsourcing of U.S. manufacturing and the decline of manufacturing education and knowledge-base in the last two decades. It aspires to reinvigorate American manufacturing by placing the latest computer-aided design and advanced manufacturing technologies in the hands of innovative individuals and organizations, with the training, support, and friendly expert consultants necessary to help turn ideas into products, and products into businesses. This product development and entrepreneurship focus informs the range of tools and services that the facility provides.

It is known that traditional prototyping and product development can be very intimidating and expensive, so this facility aspires to be a one-stop shop, where anyone can make any physical item, with any material, using next generation digital design to leverage their own creativity, automated fabrication tools to reduce cost, and a community of collaborators to overcome challenges that cannot be surmounted alone. Like other such facilities it offers both membership access and contract service work because every individual project has its own constraints of schedule, budget, and performance so this arrangement allows one to find the optimal approach each time.

Membership allows you to do it yourself when you want to learn how and have the time or when no one could achieve what you are capable of doing. Contract services allow you to hire NextFab when you need assistance or when you cannot spare the time to do the work yourself.

Chapter 6   Engender Demand

Vision without action is a daydream. Action without vision is a nightmare
- Japanese proverb

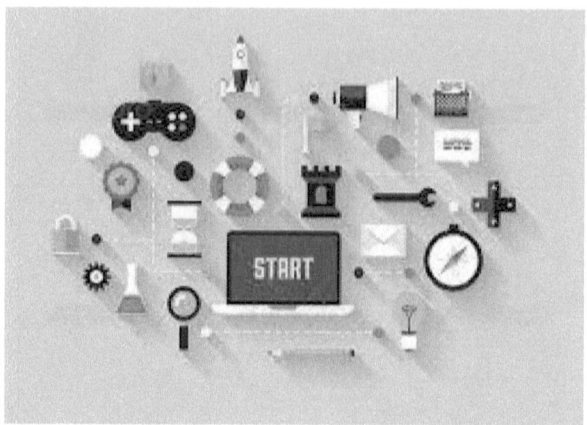

Successful market entry by startups implies engendering demand through the creation of new products and services offered. Early-stage companies offered non-dilutive funding through the U.S. federal government cover the technology spectrum from garage-to-market software-based startups to lab-to-manufacture hardware-oriented companies, and especially in the latter case towards seeding future industries whose products are manufactured in America. At this point in the technology development curve, some startups build a generic technology platform; while others may involve aggregation, clustering, and bundling; and yet others may consider pilot-plant scale-up. Some are proof-of-concept efforts, some involve the development of infra-technologies to fill-out critical gaps in the industrial commons, and some others design and build instruments and tools to advance the scientific frontier.

In addition to such financial support, further assistance can be provided that leverages public-private partnerships to advance American innovation by efficiently transforming public investments in basic research into useful products and services that create economic value and enhance American competitiveness. This chapter describes typical startup needs, the requirements of how best to facilitate their market entry and the customarily deployed market entry pathways. It addresses the question of how to build-out the required technology-specific networks that can assist startups get to market, and in specific cases, tackle scale-up issues tying startups to domestic manufacturing, local technology clusters, and supply-chains.

Such facilitation is not envisaged to be a one-size-fits-all initiative but in general to provide unstructured, highly impactful assistance. It is to help founders attract outside resources – people, knowledge, information, data, and capital. Historically, sixty-five percent of failures are attributed to problems within the startup's management team. Across entrepreneurs as a whole, wealth and control repeatedly come in conflict with each other. Few founders of high-potential startups can achieve both wealth and power; most choose between one and the other and often end up with neither[151]. Additionally, the high-tech market is a set of actual or potential customers, for a given set of products or services, who have a common set of needs or wants, and who reference each other when making a buying decision. A chasm exists between two distinct marketplaces for technology products. The first, an emerging market dominated by early adopters who are quick to appreciate the nature and benefits of the new development. The second is a mainstream market representing people who want the benefits of the new technology but who do not want to "experience" it in all its gory details. The transition between these two markets is anywhere but smooth[152].

Each technology segment has any number of specific characteristics unique to that particular sector. Almost no discovery, no general-purpose technology came from design and planning. The strategy for discoverers and entrepreneurs is to rely less on top-down planning and focus on maximum tinkering and recognizing opportunities when they present themselves. The case of lasers, a typical "solution looking for a problem", led in time to compact disks, eyesight corrections, microsurgery, and data storage and retrieval, all unforeseen applications of the technology. As so often happens in discovery, those looking for evidence did not find it; those not looking for it found it and were hailed as discoverers[153]. Another example is provided by biotechnology the core of which is not applications engineering or translational research, it is basic research, which means by its very nature that a profound and persistent uncertainty pervades the entire sector, an uncertainty that does not disappear with the next discovery, or the one after that. The biotech industry is influenced by, and faces unique pressures from scientific, legal, regulatory, political, and market factors[154].

Profiles of three startups are presented in this chapter: a software startup based in Menlo Park, in the very center of Silicon Valley, a company founded by a serial entrepreneur. It is developing next-generation automated internet search to repurpose approved drugs, one of many applications it is pursuing. The second is a company that designs and develops next-generation Federal Drug

---

[151]Wasserman, Noam - The Founder's Dilemmas: Anticipating and Avoiding the Pitfalls That Can Sink a Startup, 2012

[152]Moore, Geoffrey - Crossing the Chasm: Marketing and Selling Technology Products to Mainstream Customers, 1991

[153]Taleb, Nassim - The Black Swan: The Impact of the Highly Improbable, 2007

[154]Friedman, Yali - Building Biotechnology: Business, Regulations, Patents, Law, Politics, Science, 2008

Administration-compliant medical devices, a startup with the capability to transform a research/design concept all the way to regulatory approval. The final startup featured in this chapter can be characterized as a startup holding company that grooms small businesses performing basic research in cancer therapeutics and biologics until they are ready for large investments from institutional financiers.

Overarching Considerations

Specific issues to address and the particular market pathways to follow are technology and time-dependent. To commercialize certain inventions seems relatively straightforward, with private capital readily available for growth and profitability, while others may have regulatory barriers to overcome, or the science is difficult and time-consuming yet could lead to commercially valuable outcomes and societal impact, or require expensive infrastructure and facilities, or require long-term patient capital, or have manufacturing needs that are hard to fulfill with hollowed-out domestic supply-chains, or various combinations of these. Additionally, there are technology-specific intellectually property considerations. For example, patents in the chemical, pharmaceutical, and biotech sectors have clear-cut boundaries and are extremely valuable while it may not be so in electronics hardware where extensive cross-licensing may be required or in the case of software where ownership is often hard to discern. In other instances trade secrets and tacit knowledge may be more valuable. Intellectual property strategy and assistance provided by an outside private entity, if any, in executing this strategy will therefore have to align with such technology segment nuances.

In terms of infrastructure, facilities, investment requirements, and development time, at one end of the technology spectrum are software applications that in general have quick turn-around times, forcing startups to be agile and nimble. In such cases, there is need for flexibility, and relatively little capital is required because large expenditures in instruments and other sophisticated equipment are not involved. By the same token, in the middle of the spectrum are technologies incorporating electronic hardware such as in electro-optics, photonics, sensors, wireless, telecommunications, mobile device power management, measurement and metrology, instrumentation and control, and robotics. At the other end are the development of new materials, biotechnology, drug delivery, medical devices, and chemical technologies. Drug delivery and medical devices, for instance, may require FDA approval that could take anywhere between three to ten years. Also at this end of the spectrum are instruments and systems required for breakthroughs in fundamental sciences like nanotechnology and exotic new materials. In instances where innovations are primed for scale-up to pilot-scale production and large-volume manufacturing, tens and often hundreds of millions of dollars would be required with timeframes of a decade or two.

The four phases of a typical S-curve of a technology life-cycle are the research and development phase when incomes from required inputs are negative and prospects for failure high, the ascent phase when out-of-pocket costs have been recovered and the technology begins to gather strength, the mature phase when gain is high and stable, and the decline phase of reducing fortunes and utility of the technology. The market-oriented, grant-funding portion of the U.S. federal government's small business innovation research program is by far the largest equity-free funding mechanism in the world for early stage high-tech small businesses, those undertaking extreme technical risk. Technologies ranging from software, wireless, optoelectronics, robotics, instruments, semiconductors, energy, chemical, biotech, and medical devices are developed. Over the last three decades and more, startups funded by this program have exited, forged significant strategic partnerships, and raised substantial private financing. Such small businesses that succeed create strong strategic alliances with corporate partners and manufacturers, and seek equity financing from venture capitalists and angel investors. They have brought products and services to the market, have sometimes disrupted commerce, and helped create sustainable high-tech, high-wage jobs in America.

Startups developing new products, processes and services can generally be categorized as belonging to four basic types: Bringing a product or service into a newly created market; into an existing market; into an existing market while re-segmenting that market as a low-cost entrant; and into an existing market and re-segmenting that market as a niche entrant. A second characterization of high-tech small businesses can be:

- Build and sell a product or service
- The company is itself the innovation and the product/service so developed is the company (e.g. Google)
- Develop an intellectual property portfolio and license to someone else who then makes and sells the product or offers the service
- Use the innovation to enhance an internal process such as in semiconductor manufacturing

With regard to startup financing, it is clear from a study of U.S. venture capital investments over the last ten-year period that a skewed trend appears in the early-stage high-tech startup investment market towards information and communication technologies, with lately an additional marked inclination towards a few more cleantech investments. As Table 5.3 has shown, there are orders-of-magnitude more venture capital firms that invest in software than in hardware or manufacturing-related startups. A second attribute is the inherent nature of hard-science businesses that makes it extremely difficult to gain market traction, obtain customers, and consummate investment rounds. These often involve many hours of fruitless effort. To be able to increase the odds of success when assisting startups, we examine a number of factors beginning with the needs of a typical startup, as listed in Table 6.1. The guiding principles for such assistance should be transparency, fairness, efficient resource utilization along with robust coverage of the technology spectrum, and rapidly building credibility and trust. It is critical to maintain good working relationships with federal program managers, obtain their buy-in, and develop regular communication protocols to properly interact with them. The objective should be to deliver high-value, chaperoned introductions to potential partners, customers and investors in order to address the high-priority and tailored commercialization needs of each small business while being economical in demands on the startup's time.

Table 6.1   Typical Startup Needs

Assist with business strategy; help develop viable business models
Strengthen grantee company management team and key personnel
Assist with IP analysis/strategy; help acquire additional IP if required
Help with valuations; help raise sufficient capital as and when required
Identify and build relationships with potential customers
Identify and build relationships with strategic partners
Provide hands-on assistance with business negotiations
Help form a Board of Directors and if required a Board of Advisors
Find appropriate and reliable manufacturing facilities for startup use
Identify and leverage local entities and resources
Coach and prepare startups for investor/customer/partner meetings

A "mentor" program modeled after the Massachusetts Institute of Technology (MIT) Venture Mentoring Service (VMS) founded in 2000 can be tailored to address the needs of SBIR-funded companies. The organizational structure and rules of governance of the MIT VMS were designed by two successful MIT-affiliated serial entrepreneurs to remedy a gap in how that university supports emerging entrepreneurial ventures. The primary factor enabling this service to leverage impact and scale is volunteerism - it has almost two hundred volunteer mentors who provide startup mentoring and also assist in the operation and management of the program. These mentors contribute over twelve thousand hours of volunteer time to mentoring, program leadership and disseminating practices to other institutions annually. The program's premise is to assign a set of mentors to each startup - one mentor from the startup's region, the second with pertinent technology expertise, and the third with relevant business expertise, along with a well-defined "conflicts-of-interest" agreement.

In 2006, with the support of the Kauffman Foundation, VMS published a report and launched a workshop outlining the program and its practices to aid others in replicating this service. Beyond having a set of mentors, startups often articulate the need for hands-on assistance with identifying and obtaining potential customers, assistance with raising capital, building the team and management, hands-on help with negotiating joint development agreements, and assistance with proper valuation of their company, technology, and intellectual property for investment purposes.

Next-generation Internet Search
Automated Search to Repurpose Approved Drugs

This startup based in Menlo Park, California is being provided federal funding from December 2013 to November 2016 in the amount of approximately $975,000 for R&D work to optimize and scale a serendipitous document search system for repurposing technologies by analogy into lateral fields. By sub-parsing discrete content into ontologically separable entities such as capability, characteristic, and composition, and by comparatively assessing certain of these attributes between such entities, the attribute relatedness of these entities can be used to drive their self-assembly into related attribute networks. This approach provides a significant value proposition for drug repurposing for example. To scale the pair-wise comparison and network assembly of millions of documents, a map-reduce based text-processing framework will be developed so that massively parallel computations can be carried out in a time- and cost-efficient manner. A distributed search engine technology will be deployed to enable rapid querying of the emerging document relationship network. A series of machine learning algorithms will then be used to determine potentially hidden structural architectural features within the document relationship network. Machine learning will elucidate the nature of the relationships in drug networks through analyses of inter-node relationships and sub-graph motifs. Documents including U.S. patents and scientific papers will be processed in the system.

This effort, if successful, will accelerate the pace of R&D to enable more rapid deployment of technologies into commercial/industrial contexts. In many fields, information is expanding at such an exponential rate that finding relevant results to technical knowledge searches is increasingly difficult. Further, content is expanding so fast that most fields are rapidly forming sub-disciplines, leading to silos of different knowledge sub-domains, a clear challenge to both academia and industry. We need ever better ways to organize and present information to users. There are disadvantages of the current search engines, mostly relating to excessive similarity in search results. Further, while these engines present information relating to a known search target, they are less effective at presenting unexpected results for information that a user has never heard of but that would be useful. What is therefore needed is an exploration system giving searchers a strong serendipitous element with a maximum likelihood of results from diverse, unexpected, and potentially provocative sources. This will break down silos by providing a rapid, relevant means for knowledge-transfer between different disciplines, fostering interdisciplinary innovation. This system has been designed to provide a means for systematic, automated discovery.

Other approaches, beyond VMS, should also be tried and tested – if some of these fail to lead to timely positive outcomes they should be quickly discarded following a "fail-fast" approach. The core- or tag-team concept consisting of the entrepreneurial team, the federal program manager along with his/her network, a private entity that provides commercialization assistance, and mentors can be applied to specific technology segments, and to the benefit of a startup. Sometimes a swat team approach can be deployed when a particular startup needs quick help, and when the assistance requested is specific and well-defined. For certain difficult-to-commercialize technologies, the help of industry veterans, people who possess deep domain expertise, have taken companies public, and maintain critical industry connections, can be of great help. The above execution models can be tailored to specific technology sectors such as medical devices, robotics, cleantech, biotech, and materials.

To engage U.S. industry, large-company senior managers can be requested to volunteer to spend a few hours a month interacting with one or more startups to help guide their strategy, business model, and operations. As explained later, a web portal can be created for large companies to join based on an annual subscription fee, to allow them to more closely observe and follow startups of interest in terms of the technology development path, the attendant timelines, and probable commercial outcomes. These can in some cases lead to joint development agreements and/or acquisition opportunities. This can have great potential and if successful, be scaled up. The key to its efficacy is to develop a functional, fully-automated innovation web portal with many hundreds of featured high-tech startups and dozens of corporations closely interacting with each other.

Facilitate Market Entry

Facilitating market entry must include the capability to provide coverage of the entire technology spectrum, from information and communication technologies to chemical technologies to materials to biotech to manufacturing. In this section we discuss a set of requirements that can facilitate market entry of startup products and services emanating from innovation research-based programs, such as the need to build-out powerful networks, to develop a web portal that includes an intellectual property exchange platform, the usefulness of manufacturing partnerships, the assistance that can be provided by strategic partners, and the need for a dedicated venture capital fund.

Powerful Networks

Innovation research-based programs should have access to powerful industry sector-specific networks. Industry players; investors; researchers from universities, private and federal R&D labs; decision-makers in technology-transfer offices; licensing professionals; law firms; IP firms; past and present senior executives in industry and government (federal, state, local); supply-chain vendors; and market analysts should comprise this network. Embedded in this network should be networks of serial entrepreneurs, mentors, individual angels and angel networks, individuals with business development expertise, the limited and general partners of venture capital firms, private equity firms, investment banks, high net-worth individuals, industry trade groups, professional societies, and potential startup Board members. In some metro areas, networks of local rainmakers and "tribal" elders should be created and leveraged. The requirement is to be able to provide high-value connections to industry partners, early adopters and customers, and investors. To be able to assist in the customer discovery process, and to identify and provide access to early adopters in specific market sectors is critical. Strategies to capture early adoption can include, if appropriate, the use of the rental option and free trials. Early adopters are especially useful in the research tools and instrumentation segments, in chemical industry processes, renewable energy, semiconductors, consumer electronics, information technology, and orphan drug development.

In essence, there is a need to build-out robust networks along the ten spokes as previously enumerated in Chapter 5 – the entrepreneurs, intellectual property rights, early adopters, technology-specific networks, industry, manufacturing, investors, universities, government, and support networks. These networks are especially helpful to startups in technology areas where success is difficult and sometimes elusive because of their "hard-science" nature, and for lack of sufficient and timely support from investors and industry. Coverage provided should span the entire gamut of technologies supported such as manufacturing, advanced materials, biotechnology, medical devices, chemical and environmental technologies, cleantech, sensors, robotics, optoelectronics, semiconductors, education, and information and communication technologies. Table 6.2 is a partial compilation of the mechanisms that can be deployed to build-out the ten-spoke network for each technology segment.

Table 6.2   Network Build-out Mechanisms

Leverage existing networks of federal program officers, SBIR-funded alumni companies and current startups
Special events at university campuses
Technology cluster-driven events and workshops
Participation by portfolio companies at trade shows, technology fairs, venture forums, and professional society meetings
Feature startups at incubators and accelerators
Educate personnel at federally-funded and industry research centers
Webinars, lectures, and prize competitions
Showcase startups at innovation conferences
Discussion of startup case studies at sponsored events
Develop strategic industry partnerships
Creation of a powerful, highly influential Board

Such industry sector-specific networks should possess the capacity to evolve into a vertically-integrated innovation ecosystem, to be able to develop the assets and the means to assist all manner of SBIR-funded companies, from garage-to-market software-based small businesses to lab-to-manufacture hardware startups. It should also be well-positioned to provide coverage of the entire spectrum of technologies that populate this startup portfolio. This initiative should be built from the ground up, specifically to support commercialization of startup technologies funded by this program. The assistance should be customized to the needs of individual startups, and operate at the scale of a "cookie-cutter" program while simultaneously delivering acknowledged superior results.

This commercialization effort must lead to knowledge gained and shared, tacit and otherwise, between a facilitating private entity, hundreds of startup teams, federal program officers, and innumerable individuals that populate the many different networks created for this purpose. It should provide every aspect and every kind of essential assistance any high-tech startup might need, from providing assistance with developing viable business models; building the startup team, board, and advisors; assisting with IP strategy; raising equity capital; assisting with business negotiations; helping to acquire customers and strategic partners; finding manufacturing facilities if appropriate; leveraging local entities and resources; and preparing startups for critical meetings.

It is all about the right set of outcomes, and such outcomes require powerful networks. What startups want help with are useful and tangible outcomes. The right set of contacts tailored to individual startups can have a considerable impact on such small businesses. The introduction of potential customers and partners and investors is but the tip of the iceberg; the real work is in building-out, maintaining, and sustaining such relationships, and vetting both parties ahead of introductions. Positive impactful outcomes can be obtained, on average, for about twenty percent of the portfolio while approximately seventy percent of the portfolio may fall into the category "effort, but no outcomes", and the remaining ten percent in the "minimal opportunity for impact" category because of issues with the team and/or the lack of a proper network. The stronger this network becomes, the higher will be the percentage of positive outcomes. Additionally, knowledge of support networks in the startup's surrounding innovation ecosystem, typical elements of which were depicted in Table 5.5 can be useful in commercialization efforts.

The build-out of such industry-specific networks is especially challenging in the biotech, medical device, and pharmaceutical industries. Nonetheless, even in these sectors, much can be done such as assist with a biotech startup's initial public offering, help negotiate multi-million dollar investments from venture capital firms for biomaterials companies, assist protein-analytics startups obtain investments from the venture arms of strategic partners, and help medical device companies consummate a joint development and partner agreement with multinationals. Sustained industry and university outreach efforts and connectivity with key players and thought leaders can effectively position such startup assistance to help introduce nascent technologies harvested within this program to established large corporations and small and medium enterprises across America.

Web Portal/Intellectual Property Exchange Platform

A web portal that is also an intellectual property (IP) exchange platform is primarily meant to create better visibility to the SBIR-funded startup portfolio. Currently, portfolio information is only available through publicly available abstracts of such awards and through an online structured database, neither of which are user-friendly or in line with contemporary best practices. This portal should feature the entire portfolio accessible through Google-style, unstructured search capabilities, as well as allow interested parties to "follow" and communicate with startups. Potential strategic partners indicate that this functionality would greatly facilitate and accelerate desired interactions with startups.

The fully automated system will feature profiles of startups, sponsors, and strategic partners; facilitate the startup intake process for help with their commercialization efforts; and provide access to federal program managers. Social media aspects and featured innovation developments through technology cluster-based blogs, white papers, reports, startup narratives, press releases, repositories of video clips and audio excerpts, case studies, supply-chain maps, intellectual property maps, and a listing of manufacturing facilities can be additional attributes. Auto updates, registration, and security features should also be incorporated. The goal is to evolve into a web portal that will serve as a user-friendly national innovation resource.

The overall objective is to match startups with potential customers, investors, and strategic partners that demonstrate interest in the technologies being developed. An automated operational system is essential. The system architecture and design of this web portal should incorporate components that are public-facing, startup-facing, and program-facing. The public-facing component will provide customers, investors and corporations, access to the SBIR portfolio via social media features, security, auto-updates and Google-like keyword searches. It will be used for user registration, the development of LinkedIn-like communities, and feature market-facing profiles of portfolio companies. The startup-facing aspect is for federally-funded portfolio companies to use the portal as a web-based marketing tool capable of real-time profile updates. The program-facing component will be for internal use by federal program officers to conduct portfolio analysis; help cluster companies by technology; for general program management; to provide automatic updates of startup-related activities, news items, and technology development progress; and to track commercial outcomes.

From Concept to Regulatory Approval
Minimally Invasive Surgery Inside Active Organs

This federally-funded research and development work from April 2013 to September 2016 in the amount of approximately $1.1 million to this startup based in Bellefonte, Pennsylvania is to finalize the grip-act-reposition miniaturized stable working platform for minimally invasive procedures inside active organs, with initial focus on treating atrial fibrillation (AF), an irregular and often rapid heart rate that commonly causes poor blood flow to the body. A common procedure for treating AF is radio frequency (RF) catheter ablation in the heart. A catheter with an electrode is used to deliver RF electrical current to tissue burning paths that electrically isolate the fibrillation trigger sources. With existing catheter-based devices, ablation lines must be formed from a high number of discrete lesions created sequentially. Forming contiguous lesions in a beating heart is challenging, resulting in long procedure times and first-time success rates, for some forms of AF, as low as 30%. This technology can revolutionize catheter-based minimally invasive procedures by providing a mechanism to enter a body cavity or organ, such as the heart or lungs which move constantly, clamp to a tissue wall, and create a stable platform from which to operate and/or deliver various local treatments to tissues. The rotating gear mechanism and simple grasp-and-release function will allow repositioning within the cavity in a continuous line or arc without losing contact with the tissue wall. AF affects more than two million Americans, and approximately twenty million people worldwide. The cost to the U.S. healthcare system is almost $7 billion annually in procedures and maintenance medications. Existing catheter ablation procedures can cost over $12,000, can take over 4 hours to perform with success rates of 30-80%. Typical solutions to AF are often treatment with medications costing more than $3,000 yearly with many side-effects. This novel technique will reduce procedure time in the case of RF ablation to an hour and improve success rates, reducing the need for repeat catheter ablation procedures. Cardiac ablation will become a more viable treatment for patients on maintenance medications. Additionally, this technology can be applied in other medical applications and conditions, such as repairing difficult bleeds in the intestines, where precise navigation within a moving organ, loose tissue, or a body cavity is desired.

Currently, visibility into the SBIR portfolio is limited by cumbersome structured searches of the publicly available award database. Automating this process will lead to the creation of an online community for startups through a "microsite" for each small business. The microsite would be akin to a "profile" on LinkedIn. It would reformat information currently in the www.sbir.gov database and make it available via an unstructured search on this web portal. The search results are "microsites" - a microsite would include content about the startup, and include abstracts of their technology, videos, or even content generated by the startup itself. The web portal will make the publicly available parts of these microsites search engine-optimized.

The interface would not only allow potential partners to search for and study startups in a "self-service" mode, but also to track and communicate with startups of interest. The system will also be able to follow partner activity to allow startups to not only be informed of potential leads but also to provide reports of activity on their microsite – for example, "The following seven companies viewed your microsite this month". Potential partners will be required to self-identify and profile themselves in terms of their innovation areas of interest. This would be the beginning of an expert system matching startups to industry partners.

The portal's IP exchange platform will provide greater visibility to the intellectual property available for licensing or technology transfer from university technology licensing offices, industry, private research institutions, research institutes receiving U.S. government funds, and law firms engaged in intellectual property transactions. This IP management and licensing platform can be built around social collaborative tools to facilitate the seamless transfer of scientific knowledge and technology knowhow from research centers to high-tech small businesses and U.S. industry.

Agreements affecting intellectual property assets are critical to the underlying value of those assets. Great intellectual property can be diminished through agreements of poor quality. Likewise, the relative value of any intellectual property can be enhanced through thoughtful and effective agreements. Law firms' licensing practitioners routinely handle many different types of intellectual property licensing and technology agreements for organizations ranging from startups to technology transfer offices to multinational corporations. Some law firms offer pro bono work upfront, helping with corporate structure and IP strategy, and then later take a small equity position in the startup when it closes its first institutional investment round. About 90-95 percent of all patents sit idle on corporate shelves when companies decide not to develop them into products. Now, not-for-profit groups and state governments are asking companies to donate dormant patents so they can be passed on to local entrepreneurs who can then try to build businesses out of them.

The United States Patent and Trademark Office (USPTO) is the federal agency for granting U.S. patents and registering trademarks. In doing this, the USPTO fulfills the mandate of Article I, Section 8, Clause 8 of the Constitution that the legislative branch "promote the Progress of Science and Useful Arts, by securing for limited Times to Authors and Inventors the exclusive Right to their respective Writings and Discoveries." The USPTO registers trademarks based on the commerce clause of the Constitution - Article I, Section 8, Clause 3. Under this system of protection, American industry has flourished. New products have been invented, new uses for old ones discovered, and employment opportunities created for millions of Americans. The strength and vitality of the U.S. economy depends directly on effective mechanisms that protect new ideas and investments in innovation and creativity. The continued demand for patents and trademarks underscores the ingenuity of American inventors and entrepreneurs. The USPTO can be considered to be at the cutting edge of the nation's technological progress and achievement.

There exist various industry-based IP platforms such as the American Express IP Exchange for the financial sector, the one offered by Nike in the consumer market space, and the websites of Thomson Innovation and Intellectual Ventures. Additionally, platforms such as iBridge exclusively feature intellectual property generated by universities. The iBridge Network was created by the Kauffman Foundation and is now operated by a private entity. Globally, it is the largest and most popular online marketplace for non-profit IP. The site lists over 17,500 patents, has more than 175 participating institutions and over 13,000 members, and receives approximately 25,000 monthly hits. The site's significant traffic consists largely of corporations, startups, and venture capitalists. Intellectual Ventures believes ideas are valuable and that this firm exists to ensure a market for invention continues to thrive. The firm maintains that it has built the invention capital market from scratch and directly tackles business, problem solving, and invention. Their cross-disciplinary approach affords them the opportunity to work with leading inventors and pioneering companies to find creative solutions.

Manufacturing Partnerships

Another use-case involves startups that have manufacturing needs. Experience suggests that about a third of SBIR startups have manufacturing needs – some of these small businesses often find it difficult to fulfill these requirements in the U.S. at reasonable prices. This usually would best be provided by small- and medium-sized enterprises because larger manufacturers are, in most cases, not sufficiently interested in pre-pilot-scale volumes. It is also true that some large manufacturers with excess capacity are highly motivated to offer their underused manufacturing facilities for use by startups at subsidized rates.

We can partner with groups such as the National Association of Manufacturers, the Manufacturing Extension Partnership, and local government, business, and manufacturing consortia and make them aware of the web portal, and to help identify U.S factories with underutilized manufacturing capabilities that startups can leverage. In this instance manufacturing supply-chain maps can be developed with the intent of having U.S. manufacturers bid on startup projects that require pilot-scale manufacture as a prelude to scale-up. The portal would also allow federally-funded companies to view the profiles of manufacturers to determine if they can benefit from partnering with them in order to manufacture their product or perhaps identify a manufacturer who might be interested in incorporating their product, process, or service into their existing production lines. Startups can also possibly use crowdfunding to finance their manufacturing needs – for example Mosaic (https//:joinmosaic.com) leveraged this avenue to raise $1.1 million for home solar panel loans to finance a dozen solar projects. Additional details on how best to assist startups manufacture in America are provided in Chapter 9.

Strategic Partners

A variety of key formal and informal partnerships at the national and regional levels can be initiated, built and sustained to assist startups. These relationships can often prove to be critical to the needs of some startup companies. Table 6.3 provides a partial list of possible strategic partners.

Table 6.3   Partial List of Potential Partners

Albany Nanotech
Allen & Company
American Bankers Association
Angel Capital Association
Biotechnology Industry Organization
Consumer Electronics Association
Department of the Treasury
Eastman Business Park
Federal Laboratory Consortium
Industrial Research Institute
Jefferies Group
Kauffman Foundation
Long Island Life Sciences Initiative
MIT Deshpande Center for Technological Innovation
MIT Enterprise Forum
MIT Venture Mentoring Service
National Governors Association
National Institutes of Health
National Venture Capital Association
NIST Manufacturing Extension Partnership
Robotics Technology Consortium
Society for Industrial Microbiology
Society for the Plastics Industry
Stanford StartX
Stanford Technology Ventures Program
Technology Review
The Robotics Institute
Think Tanks – e.g. Brookings Institution; Heritage Foundation
United States Patents and Trademarks Office

Venture Capital Fund

For reasons previously alluded to, the U.S. venture capital model, in critical technology sectors, is broken. There is, for example, little access to private capital for technology startups developing hardware that require pilot-scale, and eventually large-volume manufacture of high-tech products in America. The same holds for technologies based on deep science, innovations that may often require a decade or more to establish sufficient traction in existing markets, or those that may have the potential to create entirely new markets. This would necessarily involve long-term, smart, "patient" capital which current private equity firms are unable or unwilling to provide. For many startups, the small business innovation research program is the only available option for such early-stage capital. Many technologies developed by such startup companies are in areas that today find little traction from established venture capital firms. To tackle this entrenched problem and to correct this perceived market failure, a small private venture capital fund can be created, and in time scaled a hundred-fold to create separate, perhaps independent $100 million funds, each dedicated to investing in federally-funded startups in specific technology segments, areas that are deemed critical to seeding in America whole new industries of the future, those that at this time remain inadequately funded by private capital, and require pilot-scale manufacturing. This is explored in more detail in Chapter 9.

Go-to-Market Strategies

Typical go-to-market strategies often involve industry partnerships where startups partner with large corporations via mechanisms such as joint development agreements, technology in-licensing, startup acquisition, and in some cases a larger firm taking an equity position in the startup through its in-house venture capital fund. In other situations, development of supply-chain maps can help tailor efforts to small and medium enterprises that are relatively nimble and quick to seize opportunities to incorporate new technology developed by federally-funded startups into their business units. It must be noted that Fortune 500 companies are on average risk-averse, stodgy, and bureaucratic, especially if they perceive partnering with a startup to be less than a hundred million dollar business opportunity.

It helps startups to align with industry innovation models and technology roadmaps when possible. It also benefits startups to partner with business trade groups and for this federal innovation research program to feature its startups at technology-specific industry gatherings such as the National Association of Manufacturers' roundtables; events at SEMATECH and at other manufacturing facilities; large trade shows such as the Consumer Electronics Show and the Biotechnology Industry Organization (BIO); and professional society conferences such as the ones organized by the American Institute of Chemical Engineers and the Industrial Research Institute.

Another strategy is to acquire additional intellectual property to enter the market. This can be facilitated by accessing or becoming members of various industry-based IP platforms such as the afore-mentioned American Express IP Exchange and the one operated by Nike, or by partnering with Delphion, previously owned by IBM and recently acquired by Thomson Innovation, or by accessing platforms such as iBridge for intellectual property generated by universities. It is sometimes useful to find effective ways to exploit available corporate intellectual property that has either been shelved or likely to be donated.

As was previously indicated, about a third of federally-funded startups have mass production needs and/or opportunities in the manufacturing sector. This could involve a product that requires volume manufacturing, it could perhaps be a sensor package and network to incorporate into a current manufacturing line, it could be an instrument that allows a current manufacturing process become more efficient, or it could be the deployment of a substitute biomaterial for better performance or to meet new regulatory standards. Should the startup produce the product in-house? Should the startup outsource production? Should it partner with an American or a foreign manufacturing firm? Depending on the point in the technology life cycle and the stage of technology development, does the startup have a choice between partnering with a small or medium size enterprise or a large manufacturer? What is the scale and scope of manufacturing required, is it pilot-scale production - high-value, low volume; or low-value, large volume? What if the startup has instead developed a manufacturing-related service?

Is it possible for the startup to leverage various state initiatives such as tax breaks, provisioning of land and buildings, subsidies for equipment purchases, and the easing of regulatory barriers? There are certainly opportunities for these startup companies to gainfully benefit from using numerous under-utilized manufacturing facilities in America. One can hold forums in regions around the country where manufacturing clusters in sectors that could lead to the creation of possible new industries such as bio-manufacturing, industrial microbiology, and advanced energy storage could be identified. The systematic exploration of such opportunities is further discussed in Chapter 9.

Obviously, venture financing is another option to more speedily enter the market. Startups can seal multiple financial deals using a variety of financing sources such as angel investments, acquisitions, joint development agreements, venture capital firms, and other newly emerging financial instruments being made available. Technology clustering is in specific cases a viable go-to-market strategy. Periodic in-depth analysis of the funded portfolio of companies can result in startups being clustered by synergistic technologies, for consideration by investors, potential strategic partners, and for possible roll-up opportunities where multiple startups form a small business consortia and coordinate work on different aspects and challenges encountered in similar or same technology segments, or even fuse into a single startup. Depending on their geographic location, these startup agglomerations can then form regional innovation clusters for incubation while simultaneously being able to plug into various local and state initiatives. Other go-to-market strategies are explored in the following subsections.

Incubators and Accelerators

Numerous incubators and accelerators provide rental space and rudimentary business services. Some others take a more active interest in their tenant startups, nurturing and providing business guidance, and in some cases even staking an equity position. The approaches to managing innovation by industry are exemplified by the following companies: Technology-to-Business Centers practice "outside-in" innovation for Siemens, seeking out the latest technology innovations from startup companies, individual inventors, universities and research labs. These centers use a variety of approaches for these engagements, ranging from contract work with the startup companies, to hiring the innovators, to licensing their technologies. Successful projects result in new or enhanced products, entirely new businesses, or new independent partner companies that work with Siemens.

Google X is a semi-secret research and development facility run by Google. The lab started up in 2010 with the development of a self-driving car. In 2015, Google shifted control of X Labs from Google Inc. to Alphabet Inc. under a corporate restructuring effort. Google Life Sciences, a former division of Google X, subsequently became a direct independent subsidiary of Alphabet. As previously discussed, IBM creates programs to bring external technology inside and implements an aggressive program of investing corporate venture capital in startups to extend its markets. It leverages its powerful portfolio of semiconductor patents to make it a safe foundry for new firms seeking to enter this industry. Intel leverages external technology through careful monitoring of academic research and through corporate venture capital investments in startups developing promising new technologies. Lucent takes internal knowledge out to the external market. Its New Ventures Group enables startups not ready to obtain outside venture capital to develop their ideas further within Lucent.

Startups with appealing ideas and driven employees but with no contacts, business expertise or capital can receive all those through institutions such as Techstars and Dreamit Ventures, which receive thousands of applications every year. The handful that are selected get money, advice on strategy, marketing, leadership, legal help and access to investors – in return, the nurturers receive small equity stakes and, if they have chosen the right startups and given them the right boost, a reputation that will attract further promising corporate youngsters into their orbit. Some will be able to get initial capital at effectively no cost from crowdfunding sites such as Kickstarter and Indiegogo; an enthusiastic reception can attract bigger investors - this was the route taken by Oculus VR, a virtual-reality startup acquired in 2014 by Facebook for $2 billion[155].

A few smaller venture capital firms have begun to sprout in parts of America other than San Francisco that provide small amounts of capital (around $50,000) as outright grants with no strings attached to promising startups with market-bending ideas. After a period of six months of observation, nurturing, technology development and advice on strategy, they either decide to further fund the company with an institutional round if they perceive sufficient commercial potential and early market traction or if not they simply write-off the pre-early-stage investment. In this way, these firms believe that they are able to attract better ideas and entrepreneurs in places in the country away from the customary hotbeds of innovation.

There are also a number of university incubators such as StartX and the Georgia Tech Research Corporation (GTRC). StartX is a nonprofit entity that helps to accelerate the development of Stanford University's top entrepreneurs through experiential education in technology sectors from cleantech and biotech to consumer internet and enterprise software. Its founder companies have raised funding from prominent venture capital firms, and have been acquired by companies such as Apple, LinkedIn and Dropbox. GTRC embodies Georgia Institute of Technology's founding spirit of innovation and entrepreneurship in engineering, science, and technology. By serving as the contracting entity for all sponsored research activities at Georgia Tech, it helps researchers translate their discoveries into globally competitive products and services. It also licenses all patents, software, and trade secrets created at Georgia Tech.

Y Combinator (YC), a startup factory and tech talent-spotter has come to dominate Silicon Valley's startup scene. Its four founders stumbled upon a formula that melds the best of an investment firm and a university. It helps fledgling companies refine ideas and provides sufficient training to sell their ideas to investors. Since 2005 YC has taken on batches of promising founders, and in 2015 celebrated the funding of its thousandth startup. About half of its startups have failed but eight of its firms have become what Valley folk call "unicorns", valued at $1 billion or more. Combined, the companies it has invested in are worth around $65 billion based on their most recent funding round although its share is perhaps $1-2 billion.

By producing a group of successful alumni, YC has helped popularize the idea that startups are a viable career. Aspiring entrepreneurs attend the three-month program to learn as well as for the stamp of approval and network they can claim when they leave. YC typically pays all those it admits $120,000 in return for a 7% stake. In its investments, it has benefited from the power of "network effects",

[155]The Economist, October 24th, 2015 – The Rise of Startups in Silicon Valley

the notion that a platform becomes more valuable the more people use it. This accelerator has also given startup founders more negotiating muscle with investors by acting as a sort of union to protect their rights. In 2015, YC raised a $700 million fund to invest in startups at later funding stages – this sort of investing requires different skills, and risks annoying the venture capitalists it works with, whose turf it is invading. It is beginning to invest in companies that have not typically been of interest to tech investors, including one that works on nuclear fusion. Table 6.4 is a compilation of its greatest hits[156].

Table 6.4   Y Combinator Greatest Hits

Company	Service	Market Value($B)	Join Date
Airbnb	Rents out places to stay for local hosts	25.5	2009
Dropbox	File-sharing and storage in the cloud	10.0	2007
Stripe	Software for selling from within apps	5.0	2010
Zenefits	Online Human Resources/payroll services	4.5	2013
Instacart	Grocery collection and delivery service	2.0	2012
Docker	Platform to manage distribution of software	1.1	2010

Boston's LabCentral is an incubator for biotech startups that opened in 2013, renting space to would-be entrepreneurs. In 2014, tenant startups raised $201 million in seed-stage venture capital funding, more than the entire Swiss life-science and biotech industry. To create it, Massachusetts taxpayers fronted $5 million for the project under the state's 2008 Life Sciences Act, which authorized $1 billion in investments, grants, loans and tax breaks over ten years to give the local biotech industry a lift. This incentive package has helped the state win $761M, a third of all U.S. seed-stage funding for biotech from 2009 to 2013. Nine of the world's ten biggest drug companies have opened offices or R&D labs in Massachusetts over the past few years; R&D jobs in the industry have grown more than 21% over the same period, nearly three times the U.S. rate. Using channels such as LabCentral, the state made a bet on homegrown companies; where other states have used subsidies to create tech hubs from nothing, Massachusetts focused on unlocking the commercial potential of research already being done at Harvard, MIT, and the University of Massachusetts.

If you attract enough small companies that are doing cutting-edge research, the big companies will follow. Every dollar the state spent under the program boosted tax revenue by $1.66. The clustering effect also makes it difficult for companies to leave the state as they grow. It is also easy to meet people you want to talk to, from C-suite

---

[156]The Economist, November 7th, 2015 – Y Combinator: A School for Startups

level executives to bench scientists from large pharmaceutical companies to investors that fund early-stage companies. For example, Riparian Pharmaceuticals is developing drugs to prevent the breakdown of the lining of blood vessel walls. It was founded by a professor at Harvard Medical School, and brought into LabCentral by J&J Innovations, the venture capital division of Johnson & Johnson. Ultimately, it's about density[157].

Mergers and Acquisitions

Today few industries have been shaped more by mergers and acquisitions (M&A) than pharmaceuticals. In the first ten months of 2015, mergers involving drug companies in the S&P 500 share index were worth a total of $328 billion. There is a possible bid by Pfizer to buy Allergan, a Dublin-based firm best known for its Botox anti-wrinkle treatments - if this deal is consummated, it would create the world's largest drug firm, with a stock market value of more than $300 billion. What set the pharmaceuticals apart from many other industries are the exorbitant costs and the high risks involved in developing new products. In recent years it has appeared that the return on R&D spending is dwindling, and that blockbuster drugs are getting harder to find. That has increased the attractions of buying another firm with a promising pipeline of new medicines, using mergers to sidestep the costly business of drug research entirely – its "buy, not build" strategy. Table 6.5 lists the biggest pharmaceutical M&A as of October 2015. Another trigger for deals is that many big firms are suffering from the

Table 6.5   Biggest Pharmaceutical M&A

Year Completed	Target	Acquirer	Deal Value ($B)
2000	Warner-Lambert	Pfizer	111.7
2000	SmithKline Beecham	Glaxo Wellcome	79.6
2015	Allergan	Actavis	72.7
2004	Aventis	Sanofi-Synthelabo	71.2
2009	Wyeth	Pfizer	68.3
2013	AbbVie	Shareholders	64.0
2003	Pharmacia	Pfizer	59.8
2009	Schering-Plough	Merck & Co	53.8
2009	Genentech	Roche	46.7

expiry of patents on their drugs – this allows makers of cheap copies to grab much of their business. This is another reason to buy a rival firm with promising new medicines in development. Between 2000 and 2014 about $78 billion was lost in worldwide annual sales of branded

---

[157]Bloomberg Business Week, October 19-25, 2015 – Investment made by the State of Massachusetts Pays Off

drugs whose patents had expired. It is expected that between now and 2019 the industry will suffer a further $65 billion drop in sales[158].

Industry Sector-specific

The two federally-funded startup examples in the previous section and this section pertain to the medical devices and biotech industry sectors respectively. The first company takes a concept all the way to regulatory approval in the field of surgical instruments while the second focuses on building basic research findings in biotechnology into investment-grade startups.

Use of a Facilitator

A facilitator is an entity that is the private side of a public-private partnership created to assist federally-funded startups with their commercialization efforts. In selecting startups to work with, a private entity such as this can be guided by a set of criteria:

- A driven-to-succeed, flexible, entrepreneurial team that is malleable and willing to be coached
- Business model and success metrics
- Owns intellectual property in exciting, innovative technology at intersection of diverse disciplines
- Market attractiveness: potential, size, growth, access to customers
- Possible poster child for the innovation research program - job creation potential; societal impact
- Can the facilitator help? Startup needs versus what this facilitator can provide/deliver
- Risk profile; time horizon; required additional investment
- Company location: surrounding innovation ecosystem and existing partner set
- Scaling potential and possibility of manufacturing in America

These criteria help to initiate the boarding process whereby a startup is acclimatized to the assistance program and initial expectations are agreed upon by both sides. This establishes a sound relationship between the facilitator, the startup, and the startup federal program officer to effectively work towards impactful positive outcomes. For comparison, Table 6.6 lists typical criteria used by

---

[158]The Economist, October 17th, 2015 – Merger Mania in the Pharmaceutical Industry

venture capital firms as they consider making an investment in a startup[159].

Business Model Experimentation

The "lean startup" model[160] provides a scientific approach to creating and managing startups to get a desired product to customers' hands faster. The method teaches how to drive a startup and grow a business with maximum acceleration. It is a principled approach to new

Table 6.6   Common Venture Capital Investment Criteria

Market attractiveness: size, growth, access to customers
Product differentiation: uniqueness, patents, technical edge, profit margin
Managerial capabilities: skills in marketing, management, finance, entrepreneur references
Environmental threat resistance: technology life cycle, barriers to competitive entry, insensitivity to business cycles and downside risk protection
Cash-out potential: future opportunities to realize capital gains by merger, acquisition, or public offering
Internal factors: quality of management, performance to date, funds at risk, influence of other investors, fit with investment firm's existing portfolio, monitoring costs and valuation
External factors: market size and growth, competition and barriers to entry, likelihood of customer adoption, financial market and exit conditions
Difficulty of execution: nature of product or technology, business strategy model

product development. Too many startups begin with an idea for a product that they think people want. They then spend months, sometimes years, perfecting that product without ever showing the product, even in a very rudimentary form, to the prospective customer. When they fail to reach broad uptake from customers, it is often because they never spoke to prospective customers and determined whether or not the product was useful. When customers ultimately communicate, through their indifference, that they do not care about the idea, the startup fails. Business models can be developed around a product, a service, or a process. In a typical startup, business models are one or a combination of the types shown in Table 6.7.

---

[159]Lerner, Josh - Boulevard of Broken Dreams: Why Public Efforts to Boost Entrepreneurship and Venture Capital have Failed – and What to Do About It, 2012

[160]Eric Ries - The Lean Startup: How Today's Entrepreneurs Use Continuous Innovation to Create Radically   Successful Businesses, 2011

Prototype Thinking is a method to accelerate the innovation process. It was developed by its inventor over the course of more than a decade of work at large U.S. corporations, the Unreasonable Institute, and from experience gained investing in and mentoring numerous startups. At Google X, it was the critical process that enabled the division to transform nascent ideas into products such as Google Glass, the Self-Driving Car, and Project Loon at 10-100 times normal development timelines. This inventor has also worked on large projects of global scale at Microsoft and Yahoo!, and scaled new

---

Cancer Therapeutics
Building Discoveries into Investment-grade Startups

This small business based in Dartmouth, New Hampshire builds academic and early-stage innovations into high-potential medical companies, identifying discoveries of exceptional value at the earliest stages and moving them toward the market. It partners with inventors and their institutions, providing the developmental, financial, and business acumen required to bridge discovery and profitability. With robust funding options, a diverse and high impact programmatic pipeline, and partnerships with world-class academic institutions and industry leaders, it navigates all aspects of a complex industry, accelerating science to improve human health. It is based in the Dartmouth Regional Technology Center (DRTC), beside the Dartmouth-Hitchcock Medical Center, Dartmouth Medical School, and the Norris Cotton Cancer Center. The DRTC is a private, non-profit organization formed in 2004 to assist startup businesses, providing support programs and a 55,000 square foot mixed-use technology incubator. This company operates both a clinical diagnostic lab and a fully-equipped wet lab with all common and many specialized instruments useful for biotech R&D. It believes federal funding is essential for the development of early-stage, high-risk, high-reward therapeutics — so far the company has garnered $11.2 million from the National Institutes of Health. Such funding has allowed the company to work at very early stages of development. Thanks to this funding the company was able to achieve 100% survival and 100% durable protection against tumor re-challenge in murine models of cancer and initiation of a Phase I clinical trial currently ongoing at the Dana-Farber Cancer Institute. In 2015 this firm spun out and sold OnCyte to Celyad for $490 million in cash, stock, and milestones, plus royalties on sales. OnCyte is built around a set of federally-funded cellular therapies, originally identified at Dartmouth Medical School, and built into a clinical-stage division within this startup. Celyad is now planning multiple clinical trials to advance the lead candidate into multiple sites and oncology indications. This small firm's success depends on its ability to engage academia — it currently works with over 30 academic institutions to identify, vet, and advance the most clinically promising therapeutics and diagnostics. Federal funding has allowed the company to build core competencies in key areas from pipeline management to regulatory affairs to preclinical development.

---

projects from conception to significance, thus pioneering and practicing a unique approach to rapid prototyping, visioning, and data-driven design that has allowed his team to both get new products off the ground and move large organizations at unprecedented speeds.

Because of its focus on fast, low-cost, user experimentation, Prototype Thinking is ideally suited to turning technologies into market-ready products, and for helping startups identify, manage, and eliminate risk.

Table 6.7   Typical Startup Business Models

Grow organically through revenue generation (slow growth)
Private institutional investment (fast growth)
Technology licensing model
Scale-up to manufacture in America
Outsource manufacturing to another country
Sell a product in an existing distribution channel
Sales and marketing through an in-house team
Acquisition by or merging with another firm
Strategic partnering
Providing services
Contract R&D

A company that uses this methodology drastically improves its chances of success through increasing its rate of business and product learning. This allows a small company to make more progress whatever its level of available funding. High-tech startups may find this methodology useful in their efforts to speed products to market.

Chapter 7   Consequence

Turning unemployment into entrepreneurship will create a world without poverty
– Mohammad Yunus, Nobel Laureate

One way to source knowledge-driven innovation is to start with taxpayer-funded basic research in universities. This effort often leads to publications that extend the domain of knowledge and expand the frontiers of understanding, and in some cases can result in discoveries of great importance that can perhaps change the mind and change the world; some others lead to insights and inventions. These inventions can be converted to intellectual property by way of patents and copyrights, trade secrets and trademarks, and then translated to practice through additional research and development that we term innovation research. Now that we have a set of ideas that may have commercial potential, we need capable people in the form of an entrepreneur and his/her team to convert this intellectual property to an innovative product or service. This in turn can lead to a new market; can plant seeds for a whole new industry; can create jobs, growth, profits and taxes that consequently provide resources to fund more innovation, and thus can the cycle continue.

Some startups want to change the world, while others want a lucrative exit - an initial public offering or an acquisition by a large company for cash, stock or both. Many buyers use stock for all or part of the purchase since it allows them to leverage their balance sheet and potentially keep cash for other growth initiatives. If the stock is public there may be a lockup period of some kind. If the stock is private, there may not be a path to liquidity. An "earn out" is usually tied to performance clauses to be met before subsequently obtaining cash from the deal. Different types of buyers will value the startup differently – a private equity firm looks at the ability of the company to generate cash. Strategic buyers would like to put the startup's product or service into their channel to drive revenues. To go public, a startup generally needs to have revenues of $100 million or more although there are examples in this chapter of startups doing so with considerably less revenue. Creating cash in the form of dividends for shareholders and creating net income and cash are viable options for businesses with a small shareholder base or startups that did not take a great deal of outside investment. There are instances when a larger company acquires for the team, but not the customer value that the startup may have created. It is likely a good outcome for the team, but seldom great for the investors. Regardless of the final exit option taken, a startup always needs to focus on creating customer value and profitable growth; best to focus on building a great company, not on the exit.

It is well understood that a startup is always about an idea of how to impact a customer. What is not always given credit is how startups have a far-reaching ripple effect on the socio-economic fabric of the geography in which they operate, and all the changes that some startups have brought in their cities, in their industries, and in global markets. The impact that they have created is often beyond their customers. Apart from delivering customer value, startups have a direct impact on the cities they make their homes in, like the presence of Google in Mountain View, Microsoft transformed Redmond, Infosys impacted Bangalore and Alibaba has changed Hangzhou. When these startups grew, they directly impacted the growth of their cities as well. Employment opportunities for youth increased and new employment patterns emerged. Demand and employment opportunities for engineers saw a steep rise. It fuels local economies. Local youth had new opportunities to pursue, and experienced talent started moving to these cities in pursuit of a challenging and high-growth career.

Startups or rather their founders are locals. The live locally, shop locally, and use local resources. They employ local workers. They might rent a desk or private office in a co-working space or call a local incubator home. Startups become involved in communities and take part in events, because it is necessary for the success of their business. These combined elements inject money, resources, and vibrancy back into local communities. Sure, the dream of most startups is to move to Silicon Valley or some other innovation hotspot, but startup hubs are nonetheless refueling hundreds of cities worldwide. Startups develop well-rounded individuals - anyone working for a startup would know that they, and their colleagues, are great multitaskers. They juggle numerous roles and fulfill many functions beyond what is stated on their business cards. Working for a startup means learning to do a number of things that are necessary for the development of a successful business, in an experimental and creative environment. The few people who work for them will need to complete all tasks, and so they are valuable beyond their "specialized" function.

Thousands of small companies, employing thousands of freelancers and qualified graduates, reduces unemployment levels significantly - with small businesses hiring around 30% of the U.S. workforce.

Thanks to technological advancement and this increased competition between small businesses, startups are required and are able to develop their products quicker, more efficiently, and cheaper. As a result, we are presented with cheaper, more efficient solutions to problems that streamline business and encourage market competition. The fruits of startups, born of ideas sparked by holes in the market are used by individuals and companies alike, and so the vertical economy, tying consumers to big business, is reinforced. This influx of new ideas and products presents myriad investment options for all budgets, jump-starting the injection of capital back into the market and helping re-instill economic confidence.

The first two startup examples in this chapter aim to illustrate why the software-based garage-to-market model is fundamentally different from the hardware-based lab-to-manufacture model, the latter requiring the production of something tangible. The next two small business examples feature companies, one in the robotics industry sector and the other in the biotech sector, that are now publicly listed. The chapter ends with two more startup profiles – the first an example of a future industry sector involving a firm that has a joint development agreement to develop grid-scale energy storage technology with a large utility company. The second is a startup that has positioned itself to possibly be acquired by a multinational company in the oil and gas industry.

Canonic Market Pathways

The garage-to-market pathway is highlighted by companies such as Google, Amazon, Facebook, and Microsoft, and by the startup example featured below. Although the genesis of the technology developed by this small business was technically at MIT Lincoln Labs, the point to be made is that many software companies can indeed be seeded in a "garage". In this day of the so-called "app economy", where an idea, time, a bit of funding, a few computers, and code-writing skills is often all that is required to get a business started. It does not call for large infrastructural requirements nor a suite of facilities for prototyping, pilot manufacturing and scaled-up production as would be required in the production of hardware. This is differentiated by the lab-to-manufacture model highlighted by another startup profiled next, that requires the deployment of our country's manufacturing heritage to rebuild global competitiveness, and thus to allow startups to develop their businesses from lab-scale prototype stages of innovation to later stages of commercial production.

Strategic elements of the nation's future advanced manufacturing economy, areas such as functional thin-films, energy storage, and biomaterials, all require resources beyond computers and

---

Cybersecurity Performance
Information Security Risk Rating

This federally-funded research and development work from April 2010 to September 2015 in the amount of approximately $1.0 million to a startup based in Waltham, Massachusetts was to develop an information security ratings service. When businesses connect their networks with partners or share data with them, they are often poorly informed about the potential risks they assume. Businesses have third-party relationships for a variety of operational reasons and these partnerships almost always involve sharing sensitive and confidential data such as customer information, intellectual property, and social security numbers. Businesses worry about losing data through breaches in partner networks as they face financial, legal, and regulatory consequences. Existing risk management techniques are based on annual audits and only provide a snapshot of a partner's security posture. However, new vulnerabilities are discovered every day and the industry needs a solution that enables a business to continuously monitor changing risk posture of all its partners and proactively manage assumed risks. The objective of this R&D effort was to build a scalable fully-automated ratings system. The research focuses on identifying and incorporating new data sources, improving the statistical properties of the ratings model, and making the ratings predictive of future behavior. Historically, credit scoring has been a "cost and time-saving technology" that has provided tremendous value to lenders and borrowers alike by reducing costs, predicting future performance, and improving credit accessibility and affordability. Unlike credit scoring, no industry standard scoring service exists to rate business with respect to their information security risk. With this firm's ratings service, businesses and government will have the potential to reap the same time and cost savings that lenders derive from credit scoring services. High-profile breaches that start in the vendor supply-chain are increasing the focus on third-party risk. Objective metrics to measure if a company is more or less secure today than it was yesterday are lacking. The tools at hand to measure and mitigate security risk are inadequate. Security assessments are useful, but static, subjective and limited. Audits and tests are costly and intrusive. To truly identify, quantify and mitigate security risk, organizations need a solution that is continuous, automated and provides objective, evidence-based measures of security performance.

This company's security rating platform generates objective ratings on companies' security performance. Using evidence of security outcomes from networks around the world, the startup applies sophisticated algorithms to produce daily security ratings. It gathers terabytes of data on security outcomes from sensors deployed across the globe. From this data, the company checks indicators of compromise, infected machines, improper configuration, and poor security hygiene. Its algorithms analyze the data for severity, frequency, duration, and confidence, and then map it to a company's known networks, creating an overall rating of that organization's security performance. These objective ratings, based on externally accessible data, give visibility into a firm's security posture over time. Security ratings, ranging from 250 to 900, similar to consumer credit scores, with higher ratings indicating better security postures, are updated and presented daily.

---

garage space. An experienced workforce with skills to develop and manufacture future technology products is essential. To support such new technologies, risks have to be sufficiently reduced in order that predictability of product maturity from concept to full-scale production is increased. Infrastructure for manufacturing innovation must be provided to ensure that next-generation processes and products not only will be invented in the U.S., but scaled up and manufactured in America as well. At this crucial juncture, this is both critical to the nation's economic future as well as to solving some of the world's sustainable development challenges.

To execute the lab-to-manufacture model implies deploying our country's manufacturing assets. A partnership with the Eastman Business Park (EBP) in Rochester, New York allowed this second startup to leverage the equipment, facilities and the manufacturing line available at EBP to successfully produce their product. This facility has now been developed into what is today one of the largest, most diverse industrial parks in America. It is the only one of its size that was built in a vertically integrated fashion to support Kodak's research and development and commercialization components. Products manufactured on-site incorporated a broad range of technological advancements over the last hundred years, spanning photography, motion pictures, healthcare, printing, national defense, and document imaging. Today, it is open to the next generation of entrepreneurs, innovators and employers, making its infrastructure available to help accelerate middle-stage technology companies.

The suite of test, validation, prototyping, and pilot manufacturing capabilities available at EBP are specifically suited to help accelerate commercial deployment of technologies developed by startups. EBP can provide resources to tackle our country's manufacturing challenge with an experienced workforce to train the next generation of manufacturers with skills to develop and produce future technology products. With over a century of deep technical expertise and infrastructure in the U.S., the Eastman Business Park can be viewed as a national model in providing existing assets to support new technologies from concept to full-scale production. The unique research, innovation, and skilled workforce capabilities that exist at EBP creates an infrastructure for manufacturing innovation to ensure that the next generation of processes and products are produced in America. EBP is a 1200-acre campus that includes 17 miles of railroad track, 16 million square feet of manufacturing, lab, warehouse and office space, 50 million gallons per day of industrial water supply, and self-generated utilities with a 125 megawatt power plant producing electricity, steam and chilled water.

In order to compete globally, the U.S. must nourish intellectual property beyond the lab from prototyping and proof-of-concept to commercial products, especially with a majority of "new economy" opportunities in energy, clean-tech, consumer electronics, computing, functional printing, and biotech requiring insights in materials science and chemistry, pilot testing, and infrastructure support. It is about preserving American manufacturing strength globally, about preserving a manufacturing asset and skilled workforce that would take billions of dollars to replicate, about refusing to cede whole new industries to other countries. It is about jobs, and finding a new path forward in the

---

Flexible Displays
Manufacturing Organic Electronic Devices

This federally-funded R&D effort from April 2010 to September 2015 in the amount of approximately $1.0 million to a startup based in Ithaca, New York was to develop a photoresist system that is compatible with a much wider range of materials than traditional photoresists, and more importantly allowing for the patterning of advanced semiconducting polymers and small molecules on existing photolithographic equipment. The company improved its fluorinated photoresist system by making two new materials with lower manufacturing cost and enhanced performance. Multiple approaches were taken to continuously improve the performance of these new materials. The scalability of one or both photoresist materials to large quantities was investigated by addressing the major challenges to scale-up efforts, including dealing with heat generation and finding a suitable initiator. This R&D work will enable the large-scale manufacturing of organic electronic devices by leveraging photolithographic infrastructure currently used in industry. The availability of these new photoresist materials in large quantities and consistent quality will help meet the performance and volume demands of the organic electronic industry, which is expected to grow rapidly once a scalable and high-yield manufacturing technique is available. Touchscreens have become ubiquitous. After liquid crystal displays, organic light emitting diode (OLED) displays represent the next generation of electronic displays for their thin, lightweight, wide viewing angles, large color palettes, fast refresh rates, and high contrast attributes. In addition to superior display characteristics, OLED materials are inherently flexible creating the possibility of bendable, rollable, and foldable displays. Many properties of organic electronic materials can be manipulated by changing the polymer chemical structure, including the ability to conduct electricity, emit light, and act as a transistor. More efficient OLED emitters and higher-mobility semiconductors are being made every year.

Standard photoresist chemistries are based on organic solvents and aqueous developers that are generally incompatible with OLED and organic thin-film transistor materials making standard lithographic processes unsuitable for manufacturing OLED displays and flexible organic backplanes. Thus alternative approaches are used to pattern organic semiconductors, none of which deliver the combination of resolution, registration, throughput, and yield that is offered by industry standard photolithography. Currently, OLEDs are used in the most popular smartphones, but it has been difficult for the industry to make OLED TVs at large scales for a reasonable price. This startup's technology is a fluorinated photoresist platform that enables the direct patterning of organic electronics, and other chemically sensitive materials, using standard photolithographic equipment. Its proprietary photoresist provides a solution allowing for the direct patterning of a wide range of organic electronic materials for OLED and flexible display applications. By bringing this technology to the $100 billion display industry, the company is leveraging the installed infrastructure to allow the manufacturing of advanced OLED displays quickly and without major capital investment. The startup has laboratory facilities at the Eastman Business Park, allowing it to draw upon the best talent from the display and chemicals industry to develop its products.

---

evolving "innovation economy". We need to monetize American innovation by utilizing existing assets, leveraging the existing workforce and intellectual resources, preserving capital, reducing risk and improving predictability. The lab-to-manufacture model featured above and the startup-EBP partnership that facilitated this shows how to help our nation's best and brightest technology startups become profitable, self- sustaining, job-creating enterprises. This small business along with four other federally-funded startups now have either their manufacturing facility or are fully located at the Eastman Business Park. This allows them multiple deposition capabilities, bench-to-manufacturing scales, and ready access to a knowledgeable, experienced manufacturing workforce. There are obviously many technology incubators, accelerators, and economic development entities but they are mostly upstream research/business facilities with little or no access to testing, development or pilot manufacturing capabilities. The closest comparison would be a state-funded manufacturing site in China focused on photovoltaics or some of the large German facilities such as the Bayer campus.

Probable Outcomes

In general, a startup can grow into a company; can pivot and iterate as it continues to search for a business model; can grow very slowly and barely breakeven; or can run out of money and shutdown. The startup may die a slow death, usually a result of lack of real traction in the market. Such startups cannot muster a compelling case that they have actually figured out what customers want and that they can learn from them quickly. If the startup is doing alright, just not well, it can raise another round. However, depending on whether it raised at a high valuation before, this new round could well be a down round. Sometimes it is required, in midstream, that the CEO be replaced - unfortunately, this almost never works. In general, a really good startup CEO is not going to work for a pre-existing startup that is failing. Other times cofounder dispute ensues and one or more cofounders leave. In other cases, employees and officers quit for "real" jobs in a key space at a huge salary bump up.

Table 7.1 indicates various possibilities when a startup needs to scale. It suggests that many outcomes are not of much benefit to America. This and experience with hundreds of high-tech startups with production requirements indicates the need for new thinking that calls for more flexible approaches and more independent program management practices customized to specific technology sectors in terms of funding cycles, funding amounts and timeframes, and perhaps more opportunities during the year to submit unsolicited innovative ideas to the small business innovation research program.

In terms of outcomes, the software company featured above was doing well enough that it foresaw a market need and had the resources to acquire a startup based in Portugal, for its Cyberfeed product that allows advanced security organizations to obtain an optimized stream of global events related to live and upcoming security threats. In the second example above, by establishing itself at the Eastman Business Park in Rochester, the firm paved the way for four other startups to relocate to this facility in order to leverage this its manufacturing assets for their own production needs. The two startups featured next are now publicly listed companies. The last two startups profiled in this chapter offer in turn an example of a possible acquisition, and an example of partnering with a large company via a joint development agreement.

Table 7.1   Startup Growth Scenarios

Go bankrupt
Organic growth through revenue generation
Initial Public Offering
Synergistic roll-up with another startup
Acquired by larger U.S. company
Technology licensed to foreign company
Relocates outside U.S.
Acquired by foreign company
Technology sold to foreign entity

Experience gained by startup activities at the EBP, technology roundtables at the Manufacturing Institute, and events featuring startups at leading American universities can help strengthen the required networks for such startups in several areas of the portfolio that present particularly difficult commercialization challenges, including advanced materials, manufacturing, robotics, and chemicals. This can be an important advantage, because for startups in these technology sectors, this federal program is one of the very few domestic sources of early-stage technology development capital, and in these cases the identification of an appropriate commercialization partner is especially valuable. The important issue of scale-up and pilot-plant production that ties into domestic manufacturing can be expanded by partnering with other under-utilized manufacturing facilities around the country.

It is often critical to seek partners that are most likely to actually do business with such startups. These are not necessarily the largest "brand-name" American corporations and venture capital funds. While there can be utility obtained from discussions with a large corporation even if is unlikely to consummate a transaction with a startup, one must be perspicacious, to gainfully extract potential value. Indeed, large corporations that almost certainly will not do business with such startups can actually impede their commercialization efforts, by creating a compelling distraction.

Perspectives on companies and venture capital funds inclined to do business with small businesses is informed by years of working with startups and potential partners. For those organizations, both large and small, that are inclined to engage seriously with these startups, we must invest significant time and effort to strengthen the startups' relationships with these entities, to better understand the

---

I Can Walk Again! Did You Say IPO?
Powered Knee-ankle Prosthetic System

This taxpayer-funded research and development work from August 2009 to February 2015 in the amount of approximately $2.0 million to a startup based in Richmond, California was to develop an integrated, powered knee-ankle joints in trans-femoral prostheses that use sensory information from the ground and the wearer, and to create an in-home gait training device that allows a post-stroke patient to undergo rehabilitation with little or no assistance. The hypothesis is that prosthesis with actively powered knee and ankle joints will significantly enhance the mobility of trans-femoral amputees while walking on level ground as well as stairs and slopes. The difficulty of delivering power to prosthetic systems has significantly impaired their ability to restore many locomotive functions. Limb loss is afflicting a growing number of military personnel serving in recent conflicts, as well as a far larger number of veterans from previous wars. The integrated knee-ankle prosthetic can have a direct impact on the mobility of the trans-femoral amputees and their quality of life, and most likely alleviate the long-term consequences related to musculoskeletal health.

Approximately 500,000 Americans survive a stroke each year. Miraculously, most stroke survivors can relearn skills, such as walking, that are lost when part of the brain is damaged. They can relearn walking most effectively if they are aided in making the correct motions by a machine or a physical therapist while attempting to walk. This training is expensive and requires that the patient make regular visits to a stroke center or qualified physical therapy center. Since 2005, this company has pioneered the field of robotic exoskeletons to augment human strength, endurance and mobility of soldiers and paraplegics. These powered exoskeleton bionic devices can be strapped on as wearable robots, and have a variety of applications in the medical, military, industrial, and consumer markets. It enables individuals with any amount of lower extremity weakness, including those who are paralyzed, to stand up and walk. The startup created a lightweight robotic exoskeleton which cradles a patient's lower extremities and torso, and maneuvers their rehabilitating limbs for them. This device's impact has been to move most post-stroke rehabilitation out of the clinical setting thereby reducing labor costs dramatically. The gait training exoskeletons will be wearable, unobtrusive, and allow patients to maneuver in the real world. Patients would therefore be able to wear such devices for most of the day, thus remaining mobile and gaining the therapeutic effects of physical therapy over the course of a day, rather than just a short session. Furthermore, creating such a device will also give clinicians an alternative to the wheelchair to assist patients who are unable to recover adequate mobility to normally function in their daily lives. This could potentially reduce unhealthy effects of wheelchair use for millions. The firm became a public company in 2014 and is now listed in the OTCQB. The OTCQB is the middle tier of the three marketplaces for trading over-the-counter stocks provided and operated by the OTC Markets Group. The OTCQB has replaced the Financial Industry Regulatory Authority (FINRA)-operated OTC Bulletin Board (OTCBB) as the main market for trading OTC securities that report to a U.S. regulator.

---

investment and innovation needs of such organizations and be careful to source only relevant and vetted small businesses to these partners, so that each introduction is taken seriously by the partner.
Opportunities sourced should be addressed to senior-levels in various

venture capital firms, at U.S.-based multinational corporations that are considered "brand-names", and with other midsize companies that

---

Treasures from the Ocean
Valuable Medical Resource Bottleneck

This publicly-funded research and development work from March 2009 to August 2013 in the amount of approximately $1.1 million to a startup based in Port Hueneme, California was to develop methods for the control of larval settlement, metamorphosis, and post-larval growth of megathura crenulata (keyhole limpet) to support the production of commercial quantities of keyhole limpet hemocyanin (KLH), a unique and medically valuable marine natural product. Unlike many other prospective medical products from marine organisms, KLH is already in extensive use in over twenty KLH-based therapeutic vaccine trials. Earlier funded research successfully identified a critical "cue" for settlement of M. crenulata larvae and demonstrated the feasibility of achieving long-term commercial objectives. This R&D effort translated these previous results into prototype designs for testing and optimization of systems, and diets and aquaculture methods for cultivation of age-specific developmental phases, from metamorphosis to fully developed adults for KLH production.

The elucidation of the underlying biochemical factors that promote settlement, metamorphosis and early post-larval survival of this carnivorous gastropod can add significantly to the body of scientific knowledge in this field and improve the potential for cultivation of other commercially important species with biomedical potential. The provisioning of sustainable commercial supplies of KLH for new, life-saving therapeutic vaccines for cancer, arthritis, hypertension, and other debilitating diseases, without continued dependence on the limited and threatened fishery is an additional benefit. Finally, providing regulators and resource managers the opportunity to formulate management policies to protect the wild population without imposing limitations on KLH or the KLH-based vaccines under development is important. KLH is an immune-stimulating protein widely used as an active pharmaceutical ingredient in many new immunotherapies, and as an injectable product to assess immune response. This versatile molecule can be combined with a disease-targeting agent to create a novel immunotherapy, or used alone to assess the body's immune response. KLH has a long history of safe and effective use across a range of disease indications such as cancer, inflammatory disease, Alzheimer's, and immune disorders. Since it can only be obtained from a scarce marine source, the giant keyhole limpet, this development in aquaculture science and KLH production ensures that this critical natural resource thrives. This firm has developed the proprietary ability to sustainably produce this essential molecule, and set new benchmarks for its manufacture and the protection of its natural marine source.

In 1999, a paradigm shift in drug research – toward treatment strategies that focus the body's own immune system to target disease – pointed to rising demand for KLH protein. This brought attention to an unsustainable situation: fishery and manufacturing practices threatened to deplete the scarce marine source of KLH and constrain future supply of this important pharmaceutical ingredient. The company was founded to address the need for sustainable, commercial-scale supplies of high-quality KLH. Today, this firm (OTCQB: SBOTF) is a publicly-held corporation specializing in the development, manufacturing and commercialization of KLH products and KLH-based immunotherapies.

---

transact with startups. A database of historic venture capital financings, customer relationships, and joint-development agreements involving startups would allow us to give such small businesses relevant information about "market rates" for private, early-stage transactions that is hard to find anywhere else. This knowledge has effectively helped to successfully mediate contentious negotiations, by providing both parties with a greater sense of reasonable expectations.

Experience and hindsight have illuminated mistakes that can be made by startups in the areas of strategy, communications, operations, and execution. Most often, when operational or execution failures occur, it is best to devise quick work-around solutions, remove real and perceived inefficiencies, and adopt corrective measures. It is important to communicate clearly with startup management in a timely fashion especially at critical junctures when the startup is rapidly changing and evolving, a time when feedback, management judgment and external perspectives can be useful. Ineffective communications can result in persistent gaps in internal relationship building and may lead to organizational mistrust. The stated task of helping to commercialize a widely varied set of technologies remains a difficult challenge.

Possible Impact

The technology, products, and services spilling out of high-tech startups can help fill critical holes in domestic supply-chains, and can develop required infra-technologies in various industrial sectors. In this way, along with strategic partnerships with U.S. manufacturers to scale production leading to more complete value-chains and the needed infrastructure technologies, startups can help anchor new technology products and services in America and assure that manufacturing in critical industrial sub-sectors remains in the United States. Additionally, such startups create high-tech, high-wage jobs and can possibly help seed whole new industries in America.

Develop Required Infra-technologies

As defined in Chapter 4, infra-technologies provide the basis for technical and functional interfaces between products and components that make up a system. It requires interoperability protocols to share data with multiple tiers in the supply-chain, and the measurement and test methods required to efficiently conduct research and development, control production, and execute marketplace transactions. Achieving efficiency for technologically advanced production systems requires process-control techniques, critical evaluation of engineering data to execute the process control, and methods and reference materials to calibrate complex equipment. Product acceptance testing protocols and standards and specialized facilities for determining compliance with industry standards are efforts to develop the industrial commons for future technologies. The next featured startup provides an example of such infra-technology, a hand-held measurement instrument being used in the oil and gas industry.

The "Industrial Commons" is the foundation of technical, design and operational knowledge and capabilities that is shared within an industry sector, such as R&D know-how, advanced process development and engineering skills, and manufacturing competencies

---

Soot Sensor to Micro-spectrometer
Innovative Use of Electron Spin Resonance

This federally-funded research and development work from April 2010 to September 2015 in the amount of approximately $1.14 million to a startup based in Foster City, California was to develop a soot sensor that eventually morphed into a micro-spectrometer. A microfluidic valve system that the company founder had previously developed to make a primitive sensor to measure air particulates using extremely small signal signatures led to the development of a soot sensor for which the company was awarded a federal grant. It soon became clear that this sensor could also be used to analyze properties of oil and other viscous liquids. In 2010, the company developed a micro-spectrometer for academic researchers, a low-performance, low-cost version of what they sell today. The startup cobbled together a rudimentary sensor package and began to market it to oil companies. This first-generation product was capable of analyzing the properties of crude oil catalysts and additives. In 2014, their fourth-generation product successfully concluded a field trial with their first paying customer. The proprietary micro-electron spin resonance (ESR) sensor technology allowed the company to reduce spectrometer size to one that you could hold in the palm of your hand when before such an instrument package would take up a room. Miniaturization and massive cost reductions in turn opened up entirely new applications in the oil and petrochemicals industry. Today, the startup offers advanced, cost-effective solutions in electron spin resonance for scientific research, education, and industrial applications. It manufactures magnetic resonance spectroscopy solutions based on their patented micro-ESR technology. They design and distribute the world's first and only purpose-built online ESR spectroscopy solution. This technology can be used for oil analysis and online analysis of petroleum products, lubricants, as well as many other industrial sensing applications including the detection of reactive oxygen and nitrogen species. The micro-ESR is also an excellent teaching tool for undergraduate chemistry and physics courses.

---

related to a specific technology. A supply- or value-chain is based on the idea of seeing a manufacturing or service organization as a system, made up of subsystems each with inputs, transformation processes and outputs involving the acquisition and consumption of resources - money, labor, materials, equipment, buildings, land, administration and management. Table 7.2 provides an example of such a value-chain for the biotech sector[161]. How value-chain activities in an industrial commons are carried out determines costs and affects profits.

---

[161]Tassey, Gregory – The Technology Imperative, 2009

Research shows that the majority of the employment startups generate remains as new firms age, creating a lasting impact on the economy. Conventional thinking on employment from startups is that many of the jobs they create evaporate as a high percentage of them fail only a few years later. While many new firms fail, destroying jobs,

Table 7.2   Biotechnology Value-chain

		Generic Technologies		
Science Base	*Infra-technologies*	*Product*	*Process*	*Market Products*
Cellular Biology	Bioinformatics	Anti-angiogenesis	Automated Cell-based Assays	Coagulation Inhibitors
Genomics	Biomarkers	Antisense	Cell Encapsulation	DNA Probes
Immunology	Biospectroscopy	Apoptosis		Drug Delivery
Microbiology/ Virology	Combinatorial Chemistry	Bioelectronics	Cell Culture	Inflammation Inhibitors
Molecular Biology	DNA Sequencing and Profiling	Biomaterials	DNA Arrays/ Chips	
Nanoscience		Biosensors	Fermentation	Hormone Restorations
Neuroscience	Electrophoresis	Functional Genomics	Gene Expression Profiling	mRNA Inhibitors
Pharmacology	Fluorescence	Gene Delivery Systems		Nanodevices
Physiology	Gene Expression		Gene Transfer	
	Gene Typing	Gene Testing	Immunoassays	Neuroactive Steroids
	Magnetic Resonance Spectroscopy	Gene Therapy	Implantable Delivery Systems	Neuro-transmitter Inhibitors
	Mass Spectroscopy	Gene Expression Systems		
	Nucleic Acid Diagnostics	High-content Screening	Noninvasive Imaging	Protease Inhibitors
	Protein Structure	Monoclonal Antibodies	Nucleic Acid Amplification	Vaccines
	Modeling/ Analysis	Pharmaco-genomics	Recombinant DNA	
		Proteomics	Separation Technologies	
		Stem-cell		
		Structural Drug Design	Transgenic Animals	
		Tissue Engineering		

many others also thrive and create jobs. This growth in employment partially balances out the jobs lost by closing and shrinking firms. The jobs created when startups are established do not disappear overnight. In fact, they are remarkably durable. When a given group of startups reaches age five, the group's employment level is eighty percent of what it was when it began. Twenty-five years after firms start, only about twenty percent of them still exist, but the employment numbers appear to level off at around sixty-eight percent of their initial values.

The fact that firms have decreased so rapidly yet employment has more or less leveled out means surviving firms continue to grow. Firms fail, but growth, even at these well-established firms, continues, keeping employment from dropping with the number of establishments.

New Industry Creation

An example of startups creating whole new industries is provided by information and communication technologies – e.g. computers, internet, consumer electronics; and more recently the biotechnology sector and the upcoming synthetic biology industry. Table 7.3 lists a possible set of platform technologies (see also Tables 8.1 and 9.6) that have the potential to create future industries in America.

Table 7.3   Possible Set of Platform Technologies

3D printing
Advanced materials
Biomimetics
Cleantech/renewable energy
Climate change mitigation
Data mining
Drones
Energy storage
Genomics and proteomics
Human-computer interfaces
Industrial-scale microbiology
Massive open online courses
Nanoscale measurement and instrumentation
Nuclear fusion
Personalized medicine
Quantum computing
Regenerative medicine
Robot sense, motion, thought, emotion
Self-driving vehicles
Semiconductors - flexible, organic, biologic
Solid-state lighting
Stem-cell based cures
Synthetic biology
Water purification and desalination

With the rise of alternate, renewable energies but intermittent sources like solar and wind power to mitigate the deleterious effects of climate change, energy storage technologies, at the device level and at electric grid-scales, are becoming increasingly important and critical, and can possibly be the start of a whole new industry. The publicly-funded startup profiled next is an example of developing a low-cost energy storage system for the large-scale electric grid in America. This company is now a global provider of grid-scale energy storage solutions for supporting a cleaner and more efficient electric grid helped along with government funding and venture capital. This startup has been granted multiple U.S. patents on core aspects of their technology, solidifying their leading position in the field of compressed-

---

For a Cleaner, More Efficient Electric Grid
Low-cost Energy Storage

This startup based in West Lebanon, New Hampshire was provided federal funding from August 2009 to July 2011 in the amount of approximately $1.0 million for R&D work to develop and evaluate a beta prototype 50 kilowatt, 300 kilowatt-hour novel energy storage system. Currently, no energy storage technology can provide low specific cost, high energy density, and long lifetime operation in the mid-capacity range of100 kilowatt-hours to 10 megawatt-hours. Particular opportunities exist with renewable energy providers, commercial and industrial consumers, as well as utilities for services such as capacity firming, consumption smoothing, energy arbitrage, and power regulation. This company intends to develop, manufacture and market a disruptive, cost-effective and scalable energy storage device that will serve as an enabling technology for the proliferation of alternative energy generation sources such as wind and solar power.

Low-cost, long-lifetime energy storage has the potential to positively impact grid stability and reliability, and reduce commercial and industrial consumer energy costs. It will likely increase the overall market penetration of renewable energy that is intermittent by nature. As the penetration of wind energy increases due to regulations requiring lower carbon footprints, rising fossil fuel costs, public sentiment surrounding climate change and other market pressures, the industry faces a number of challenges. These include intermittency of supply, errors in forecasting power production, increased regulation requirements, and transmission congestion. As wind becomes a larger percentage of the energy portfolio, these challenges are compounded. Energy storage has the potential to mitigate these integration issues, thereby allowing wind power to be utilized on a larger scale in a more economically-viable fashion. Specifically, wind generators will be able to use this startup's energy storage technology to firm capacity, increase peak sales, and enhance ancillary service capabilities. Storage capacity therefore not only represents a significant value to current wind production, but should also be seen as an enabling technology that will allow the world to reach its near- and long-term energy management goals.

---

air energy storage. The company is on track to demonstrate the first multi-megawatt, grid-connected energy storage system. Its technology enables a site-anywhere, zero-emissions storage solution through isothermal compression that allows air to reach high pressures without the inherent challenges of high temperatures or high thermal losses. Compressed air at near-ambient temperature can be stored until needed with minimal energy losses. When power is needed, isothermal expansion can deliver electrical energy with no requirement for natural gas combustion.

Part III

New Industry Creation

The rumors of the demise of the U.S. manufacturing industry are greatly exaggerated
– Elon Musk

To live well, a nation must produce well. A country that no longer makes things will eventually forget how to invent them. When we lose our ability to manufacture, we lose our ability to innovate. Manufacturing capabilities are hard to acquire and easy to destroy. Once-dominant and important U.S. industries like machine tools, auto parts, semiconductors, printed circuit boards, consumer electronics, appliances, furniture, clothing, telecom equipment, home furnishing and many others have collapsed. The fact that manufacturing is unlikely to drive significant job creation does not imply that it is irrelevant. Citing old stooges of corporate welfare, picking winners and losers, and protectionism, apostles of denial downplay key economic facts such as the growing and massive asymmetrical U.S. trade imbalance in this important technology sector[162].

The senseless debate about the virtues of free markets versus government intervention, and the collective repudiation by America's economic elite of the need for an industrial base has led our country to the precipice. History shows the U.S. economy to be one of the most market-oriented economies in the world where government policies have always played a vital complementary role in fueling economic growth. We subsidize many service sectors of the economy, employer-sponsored healthcare plans being one example, and cannot ignore the continuing decades-long massive subsidies for agriculture. Manufacturing is intimately connected to our service economy whether it is software design or basic research. When manufacturing goes overseas, design follows, when in many instances a company cannot

---

[162]Tassey, Gregory - Rationales and Mechanisms for Revitalizing U.S. Manufacturing R&D Strategies, 2010

afford to separate innovation from manufacturing. When manufacturing exits research and development funding dwindles in direct response. A vital source of competitive advantage is the industrial commons - shared sets of technical and operational capabilities, research and development (R&D), manufacturing infrastructure, knowhow, process-development skills, and engineering capabilities, some of which are shared across firms and even across industries. The collapse or hollowing-out of these commons cannot easily be recreated.

Manufacturers are painfully aware of the many short-term challenges they face. Increasing competition, volatile energy and input costs, new technologies, and supply-chain visibility are all creating immediate challenges for organizations that simultaneously are fighting to prepare for the launch of the "next wave" of innovations. Manufacturers are increasingly innovation-led and focused on improving research and development efficiency and value. Half of all manufacturers say their strategic focus is innovation-led. Thirty-two percent cite the development of new products and research and development as a top strategic priority. Thirty percent say the biggest challenge for their organization is research and development inefficiency. Sales growth and cost reductions continue to top the agenda as manufacturers prepare for increased competition. Fifty-five percent cite sales growth as a top priority and forty-one percent say they are focused on reducing the cost structure. Almost forty percent say that their biggest challenge stems from intense competition and pressure on prices.

Manufacturers are increasingly looking for breakthrough innovations and are increasing investment in R&D. Forty-one percent say their primary strategy for innovation is to pursue breakthrough advances. Seventy-four percent maintain they will spend upwards of four percent of revenues on R&D over the next two years. Manufacturers are entering into partnerships and adopting new technologies in order to improve speed-to-market and lower innovation costs. More than three-quarters insist that partnerships will form the basis of innovation for their company. Almost half say they are adopting new manufacturing technologies to drive innovation.

Reducing costs and preparing for new product launches are high priorities for manufacturing supply-chain organizations. Lowering costs and working capital levels is cited by forty-six percent as a top supply-chain priority. Twenty-nine percent state they will restructure to support growth and thirty-two percent say they are reconsidering their global footprint based on growth expectations. Concerns about supplier performance and capacity remain high but visibility into supplier organizations remains surprisingly low. Behind the need for greater flexibility, supplier performance and supplier capacity is cited as the second and third biggest supply-chain challenges globally. Yet just fourteen percent claim to have complete supplier visibility into Tier 1, 2, and beyond.

The top strategic priorities are reducing cost structure, greater speed-to-market, increasing cash flow from operations, sales growth, development of new products, and reducing operational complexity. The biggest challenges are efficiency in research and product development; managing geopolitical risk, intense competition and pressure on prices; keeping the business model competitive; and information technology systems keeping pace with demand from the business. Innovation waits for no one; those who fail to embrace the new reality of the innovation cycle will quickly be left behind. Collaboration for innovation takes many forms. Some are partnering with their suppliers and vendors to develop new innovations at the parts-level, while others are joining up with nontraditional players and technology vendors to identify, develop and commercialize new innovations[163].

In 2014 the world's manufacturers spent about $1.4 trillion on developing new products, the combined annual sales of the world's manufacturers were thirty-five trillion dollars, and unique product types were about ten billion. The U.S. still makes products but mostly in areas subject to strict government regulations and that receive heavy federal R&D investment like pharmaceuticals, medical equipment, and military weaponry or they are consumables like toiletries, processed foods, and large pieces of capital equipment. To reduce the trade deficit, we need to expand value-added manufacturing activity in the U.S. in addition to growing our service sector. Manufacturing has become knowledge work. American factories produce sophisticated goods such as biotech drugs, aircraft engines, semiconductors, specialty materials, medical devices, and scientific instruments. The post-industrial manufacturing ecosystem represents a complex and highly integrated globalized value web. This web includes cutting-edge science and technology, innovation, talent, sustainable design, systems engineering, supply-chain excellence and a wide range of smart services, as well as energy-efficient, sustainable, and low-carbon manufacturing[164].

Advanced manufacturing is the interface between the innovation system and industrial production, the creation of sustainable capabilities to make successive generations of integrated solutions coupling production of physical artifacts with services and software, increasingly drawing upon custom-designed and recycled materials enabled by the physical and biological sciences and involving engineering-manufacturing collaboration[165]. The internet is

---

[163]KPMG Global Manufacturing Outlook – Preparing for Battle: Manufacturers get Ready for Transformation, 2015

[164]Prestowitz, Clyde - The Betrayal of American Prosperity: Free Market Delusions, America's Decline, and How We Must Compete in the Post-Dollar Era, 2010

[165]Berger, Suzanne – Making in America: From Innovation to Market, The MIT Task Force on Production in the Innovation Economy, 2013

democratizing the tools both of invention and of production and the creativity of entrepreneurs and individual innovators are reinventing manufacturing, creating jobs along the way. Entrepreneurs and inventors are no longer at the mercy of large companies to manufacture their ideas thanks to 3D printers, open-hardware, and the maker movement which is now doing for physical goods what open-source did for software. Global manufacturing will soon work at any scale, from units of one to millions. Customization and small batches are no longer impossible, in fact, they are the future. Manufacturing will soon become just another "cloud service" that you can access from web browsers, and global supply-chains have become scale-free, able to serve the small as well as the large[166].

The global competitive landscape for manufacturing is undergoing a transformational shift that will reshape the drivers of economic growth, wealth creation, prosperity, and national security. Manufacturing is and will continue to be an essential path for attracting investments and spurring innovation. The competitiveness of a country's manufacturing sector is critical to its long-term economic prosperity and growth. It originates a sustainable economic system, encourages domestic and foreign investment, and improves a country's balance of payments. It creates high-value jobs, not just within the sector but spilling over into such areas as financial services, infrastructure development and maintenance, customer support, logistics, information systems, healthcare, education and training, and real estate.

A strong manufacturing sector boosts a country's intellectual capital and innovativeness, underwriting R&D, pushing the technological envelope, and driving the growth in demand for highly skilled workers and scientists. With manufacturing playing such a vital role in the economic health of the U.S., the country must in turn play a key role in building an environment in which manufacturing can thrive. Especially today, when the landscape of manufacturing dominance is shifting, synchronizing government policy with the investment decisions of manufacturing executives is critical for a nation to remain competitive and create a positive cycle of prosperity. Manufacturers now have the ability to locate anywhere in the world they believe will help them achieve a competitive advantage and best serve customers. And once business investments are made, with brick and mortar in place, they are difficult to unwind, even as circumstances change.

Manufacturers must learn to behave more like technology firms. As the "internet of things" spreads to the factory floor products are being packed with ever more sensors and connected to the internet. That is transforming manufacturing, the first shift being from products to services. By one estimate the number of wirelessly

---

[166]Anderson, Chris – Makers: The New Industrial Revolution, 2012

connected products in existence, excluding smartphones or computers, will rise from five billion today to twenty-one billion by 2020. The data these products generate are the raw material for new services such as for example machines that can order a new spare part when it is needed. Such services will often be more profitable than the products they are based on. The second, related change is the race to develop "platforms" or operating systems, a software foundation upon which lots of services and applications can be built. In industry, the idea of platforms is new, and menacing. If Apple or Google were for example to control entertainment systems in cars, and the data they throw off, many carmakers would risk becoming the computer-makers of the road: churning out bits of metal while others grab the really valuable parts. For clues on how to embrace the industrial internet of things, Germany offers early lessons. The country is an industrial powerhouse where manufacturers account for twenty-two percent of gross domestic product, compared with twelve percent in America. In 2011, Germany launched a government initiative to promote computerization of manufacturing. Several companies have launched software platforms. For many manufacturers around the world the principal sticking-point in making the digital leap is often cultural meaning that they have to forge a more open type of relationship with competitors and become less hierarchical and more entrepreneurial[167].

The internet of things will strengthen manufacturers' hands in the battle for customer loyalty. Machines will be able to tell their owners how best to dispose them off at the end of their lives. Manufacturers are realizing that the best way to sell their products is to forge personal relations with customers rather than to spend large sums on broad-brush marketing. As it becomes cheaper to add sensors and microchips to products, and to connect them to the internet, their manufacturers will know much more about how end-customers are using them. This will help them develop their products more rapidly, fix any faults more quickly and tailor products more snugly to an individual buyer's needs. General Electric, for example, uses sensors to monitor how its jet engines are performing in the air, and to diagnose emerging problems.

The old form of capitalism based on built-in obsolescence is giving way to a new one in which products get better after they are bought. Incumbent manufacturers will need to hire more information-technology specialists, who may not fit easily into a culture dominated by mechanical and electrical engineers. They will have to rethink their core competencies: for example, instead of outsourcing their data management to IT firms, they may find that the ability to crunch data

[167]The Economist, November 21st, 2015 — The Industrial Internet of Things: Machine Learning

about their products in-house is as valuable as making the products themselves. The average manufacturing business is far behind in understanding this – only nineteen percent were planning radical changes to harness the potential of smart things; and only thirty-nine percent had introduced training in digital skills[168]. The rules in many industries, from construction equipment to cars, are changing: making things matters less and knowing things more. In many cases successful companies will no longer be the ones that make the best products, but the ones that gather the best data generated by connected devices and other information and combine them to offer the best digital services, and make money with new business models.

State of American Manufacturing

Beginning in the 1980s our nation began to divorce innovation from production. High-tech manufacturers, especially in technology fields we invented and were leaders in, fled offshore, taking with them jobs as well as R&D skills needed at home[169]. There is dawning realization that high-tech manufacturing is as essential as high-tech innovation to America's economic future and the continued survival of our once-great middle class[170]. Technological innovations are inextricably linked to high-value manufacturing of new products and services. This not only provided well-paying jobs for manufacturing workers but also served as a powerful economic force multiplier that increased purchasing power for the whole community and for every position on the factory floor created as many as fifteen jobs outside manufacturing, in skilled trades, engineering, product design, transport and supply, and many service sectors.

U.S. manufacturing is responsible for nine percent of employment even though the labor content in manufactured goods is declining dramatically. In 1970 it was around twenty-five percent; today it is about four percent. Manufacturing is twelve percent of the gross domestic product, sixty percent of exports, sixty-nine percent of the $250 billion in private R&D spending, and almost ninety percent of U.S. patents involve a manufacturing component. Small and medium enterprises create about half of all manufacturing jobs and make up more than ninety percent of the nation's three hundred and thirty thousand manufacturers. Productivity growth is higher in manufacturing than in other sectors of the economy[171]. Due largely to outstanding

[168]The Economist, November 21st, 2015 – Smart Products, Smart Makers

[169]McCormack, Richard - The Plight of U.S. Manufacturing, 2009

[170]Nothhaft, Henry and Kline, David - Great Again: Revitalizing America's Entrepreneurial Leadership, 2011

[171]Ignite 1.0: Voice of American CEOs on Manufacturing Competitiveness, Council on Competitiveness, 2011

productivity growth, the prices of manufactured goods have declined since 1995 in contrast to inflation in most other sectors, with the result that manufacturers contribute to a higher standard of living for U.S. consumers. The middle class in China, India, and Brazil are growing at a staggering rate, and will buy more and more products. Should these products not be made in America? These countries see manufacturing as keys to economic success[172]. The scaling process is barely happening here anymore - ideas born here are built elsewhere. Abandoning today's commodity manufacturing locks us out of tomorrow's emerging industries.

The clean energy industry will transform economies; it offers the best possible opportunity for growth in the U.S. manufacturing sector but we are quickly forfeiting it. For example, the advanced battery market was projected to grow to $25 billion in 2015 but we lack the proper infrastructure and industrial commons for it. Only one in the top ten solar producers and only one in the top five wind turbine manufacturers are located in America. China is investing ten times more in renewable energy than the U.S. is as a percentage of gross domestic product. In solar cells, a device that was invented here, China dominates the production while the U.S. produces only five percent of the world's supply.

It is a fallacy that America should simply specialize in research and innovation and let other nations manufacture. When manufacturing is offshored innovation and the social wealth it creates inevitably follow. In fact, along with our manufacturing, our R&D and services are being outsourced overseas as well. The business strategy to focus on core competencies and outsource everything else has been the majority view among economists for some time now. These can have knock-on effects that damage the industry and the economy as a whole[173]. The industrial commons, a platform for growth, which includes R&D knowhow, advanced process engineering skills, and manufacturing competencies of major technology firms as well as their suppliers can be damaged and eventually destroyed by unchecked outsourcing. As a result the U.S. has lost knowledge, skilled people, and the supplier infrastructure needed to manufacture many of the cutting-edge products it invented. When a commons erodes, it means that the foundation upon which future innovative sectors can be built is crumbling. Its loss may cut off future opportunities for the emergence of new innovative sectors if they require close access to the same capabilities. Additionally, the rise of a commons can support the creation and growth of numerous industries.

---

[172]Liveris, Andrew – Make It in America: The Case for Re-Inventing the Economy, 2011

[173]Pisano, Gary and Shih, Willy - Does U.S. Really Need Manufacturing, Harvard Business Review Special Report, 2012

In the last ten years more than a third of America's largest factories have shut down. This has gutted our ability to produce the most advanced high-technology products and energy systems of tomorrow such as electric cars, solar cells, cellphones, wireless systems, cameras, computers, power tools, and next-generation lighting. Semiconductors are at the top of the electronics industry pyramid and knowledge spillovers that semiconductor manufacturing generates lead to much higher rates of innovation and economic growth in most advanced industries. From flat panel screens to robotics to lithium batteries, the production, jobs, and economic benefits from these breakthroughs ended up overseas. The lack of a supplier infrastructure in key technology areas of the twenty-first century such as rechargeable batteries, light-emitting diode manufacturing for display and energy-efficient lighting, semiconductor manufacturing, liquid crystal displays, and precision glass, is fast becoming an enormous hurdle.

Most innovative smaller companies are more directly dependent on the manufacturing support base for their success, and are less likely to have their own in-house machine-shop, mold or printed circuit board (PCB) ability for prototyping. Without a PCB industry, for instance, a country cannot expect to have an industrial foundation for high-technology innovation. The domestic PCB industry has shrunk to almost twenty-five percent of its size between 2000 and 2013, a period during which this industry was growing globally. In addition, the U.S. has been purchasing most of its manufacturing tools from other countries. The near total elimination of America's machine tool industry, the backbone of an industrial economy and the means by which all products are manufactured, is a dangerous development. Having tools to make things determines what can be made. The demise of a substantial domestic high-tech manufacturing sector would greatly diminish the efficiency of the U.S. innovation infrastructure[174]. In most high-tech industries product and process innovation are intertwined. Once manufacturing is outsourced process engineering expertise cannot be maintained since it depends on daily interactions with manufacturing. Without process engineering expertise companies find it increasingly difficult to conduct advanced research on next-generation process technologies. Without the ability to develop such new processes they can no longer develop new products. A society that lacks infrastructure for advanced process engineering and manufacturing will lose its ability to innovate.

Activities that generate the most value per worker such as R&D, sophisticated manufacturing, and skill-intensive traded services, are most desirable. These not only support attractive wages but also often lead to follow-on investments as well as technology and skill

---

[174]Teixeira, Thales et al - Reinventing America: Why the World Needs the U.S. to Bounce Back, Harvard Business Review Special Report, 2012

spillovers to other parts of the local economy. Too many U.S. firms base decisions about sourcing manufacturing largely on narrow financial criteria, not taking into account the potential strategic value of domestic locations. For U.S. manufacturers the development of skills is no longer the responsibility of companies. Workers are now responsible for improving their skills or learning new ones in demand. Instead of being seen as an important input that improves productivity and flexibility, training is now an overhead cost that needs to be eliminated. The ability to interface with sophisticated machines is perhaps the most important skill for a person seeking a job at a twenty-first century manufacturing plant.

Federal Role

Some believe that the market determines demand for industries; thus if there is no demand for manufacturing within the U.S. this sector should shrivel and disappear. They argue that deindustrialization and disinvestment allow for mutation of assets and individuals from older industries to more productive and efficient uses. Deindustrialization is not necessarily natural rather it is a result of a complicated set of factors - technology change, downsizing, globalization, offshoring, and deregulation. Technology change, for example, not only allows for more efficient production, thus reducing the demand for workers, but also makes movement of money, material goods, communications, and management easier over greater distances. Deindustrialization is largely controlled by corporations to enhance shareholder value.

The economic shifts in the last thirty-five years are a direct result of decisions made by corporations and government leaders to pursue economic profit rather than the good of either communities or the environment. Policymakers at the national level cannot seem to get past vapid industrial policy debates, but regional and local leaders heartily embrace industrial policy. It is proper to oppose heavy-handed industrial policy that calls for government to pick winners or when it tries to prop up domestic companies through subsidies or other targeted support. We should instead build capabilities through research in manufacturing sciences by advancing new technologies, supporting shared infrastructure, rethinking the manufacturing process, helping small- and medium-sized manufacturers with resources to stay abreast of innovative technologies and processes, and supporting the creation of industry-led technology consortia of industry-led technology consortia[175].

Government policies should focus on manufacturing capabilities that pertain to immature or newly emerging process

---

[175]Atkinson, Robert - Why We Need a National Manufacturing Technology Strategy, 2011

technologies, and those in which manufacturing-process innovation is highly interdependent with product R&D. In both cases the manufacturing capabilities need to be geographically close to research and development[176]. Economic policy should be redirected at innovative small and medium enterprises. Too much focus is directed at large manufacturers that have shifted production offshore and show scarce allegiance to local communities. Individual companies cannot justify the investment required to fully develop important new technologies or to create the full infrastructure to support world-beating advanced manufacturing technology capabilities in strategic, targeted areas.

The U.S. lags behind in its manufacturing sector innovations relative to high-wage nations such as Germany and Japan. It ranks fourth in global manufacturing competitiveness behind China, India, and South Korea. The top driver in the competitiveness index is talent-driven innovation. Business leaders in the manufacturing sector criticize U.S. immigration policy that limits the number of researchers, scientists, engineers, and skilled workers required to help them compete. The other important way government can encourage domestic manufacturing is by supporting training[177]. Business executives further stress that government policies critically affect manufacturing competitiveness, and a country's ability to compete in international markets. America now has the second-highest corporate tax rate among its major trading partners, trailing only slightly behind Japan. Corporate income taxes distort business decision making, discourage capital investment, reduce hiring, and cause firms to invest billions of dollars in tax planning, compliance, and dispute resolution that could otherwise be put to more productive uses. If manufacturing profits were taxed at fifteen percent, the same as on carried interest for private equity, then manufacturing in the U.S. would be a lot more attractive to companies around the world. Smart consistent regulations and a permanent R&D tax credit will make an impact. The annual cost of tort claims and attendant litigation exceeds $250 billion. Manufacturers believe it will become increasingly difficult for them to compete with companies in countries with lower healthcare costs. Manufacturers cannot go it alone. Government must play its part by developing policy and national manufacturing strategies that are collaborative, integrated, focused, and effective[178].

---

[176]Shih, Willy and Pisano, Gary – Producing Prosperity: Why America Needs a Manufacturing Renaissance, 2012

[177]Global Manufacturing Competitiveness Index, Deloitte, 2013

[178]Note: The six startups presented in this chapter highlight manufacturing technologies that, in the first two examples, were created in the U.S. but now are produced or will be produced abroad. The next two examples are technologies being created in the U.S. and potentially can be manufactured here provided the needed industrial commons is not allowed to wither. The last two startup examples highlight innovative, advanced manufacturing technologies being developed and deployed in America.

Traditional mechanisms by which technology is first commercialized in the U.S. is by large, vertically-integrated enterprises, and more recently when startup firms invest in promising technology from research labs, larger companies then step in to provide later-stage product development funding and market access. The current model featuring small business and venture capitalists is under stress

---

The Next Dimension in Touch
ForceTouch Stylus Experience

This taxpayer-funded research and development work from April 2014 to September 2016 in the amount of approximately $590,000 to a startup based in Atlanta, Georgia is to commercialize a force-sensitive solution for touch applications that will overcome the technical shortcomings of currently available technologies. The design utilizes the sensitivity, size, and cost advantages of microelectromechanical systems in a novel configuration to prevent overloading of the sensor for large applied forces. The solution will enable truly force-sensitive touch that is low-cost and highly sensitive. It will be operable with any object, including fingernails, gloves, and styluses, while not being susceptible to environmental factors such as dirt and moisture that hinder current capacitive technologies. The startup's mission is to bring 3D touch solutions to market and thus enable the next generation of intuitive user interfaces. It has developed force-sensitive touch technology, a third dimension of touch that enables a richer experience with mobile devices in the consumer, wearables, internet-of-things, automotive, industrial, and medical markets. Current investors include Intel Capital, Takata, and Flex. The impact of this project lies in its opportunity to disrupt the status quo in touch-based human interface technology. Two decades ago, touch technology was primarily found within ATMs and point-of-sale systems. More recently, the technology migrated to other electronic industry verticals, including one of the largest and fastest growing - smartphones and tablet PCs. Consumer demand is driving an explosion of applications in every vertical. There is increasing demand for low-cost, low-power, more feature-rich touch solutions. In addition, new user experience benefits such as force sensitivity are constantly being pursued, yet there is no viable solution yet on the market. Such a technology would be poised to capture significant market share from existing technologies. After considering the United States, Germany, China, and Taiwan, this small business is forced to choose Taiwan for its high-volume manufacturing needs.

---

are no longer efficacious. U.S. companies seldom invest in technology more than two years down, and venture capitalists more than five years out. Virtually all the latest semiconductor and display manufacturing capacity, and the infrastructure to manufacture prototypes is now located offshore. The same holds for nanotechnology which will require long-term patient financing before new products begin to transform virtually every modern industry sector.

There has been a significant deterioration of companies that design and make discrete components triggering a fundamental hollowing out of the national innovation system. A new generation of public-private partnerships dedicated to technology transition, and involving large groups of research institutions, consortia of small and large technology companies, and local economic development organizations nationwide must work together to avert wholesale loss of technology and industrial leadership in critical technology areas such as energy storage and conversion, cleantech and alternate energy,

---

Mobile Electronic Devices
Smaller, Thinner Chips for Power Management

This federally-funded R&D effort from April 2014 to September 2016 in the amount of approximately $1.4 million to this startup based in Cambridge, Massachusetts is to develop integrated voltage regulators (IVR) that are ten times smaller than existing solutions. Power management integrated circuits (PMICs) are chips in mobile electronic devices that deliver power from the battery to different processor and memory segments. Existing PMICs use discrete printed circuit board-level inductors to efficiently deliver power, but the problem is that inductors are prone to occupying large areas on the board. This is especially problematic in space-constrained mobile electronic devices. Mobile device manufacturers expend tremendous efforts to reduce square millimeters of board area to make room for additional semiconductor chips for new features. Integrated voltage regulators will allow mobile devices to have the required extra space that traditionally is dedicated to bulky discrete components for PMICs. This additional space can be used to either include more chips to provide new features or increase the battery size for longer battery life. This R&D effort will first develop semiconductor test-chip prototypes that prove the idea of IVRs, and then demonstrate the same in commercial mobile electronic devices. The goal is to ship IVRs in high-volume smartphones and tablets by replacing existing PMICs.

If successful, this work will save board area and processor power consumption in mobile devices, both of which are crucial for continued innovation in such devices. Mobile device manufacturers constantly are on the lookout for new features to add in their next generation products in order to differentiate themselves from competitors. New features often need additional chips but the constraint remains that there is simply no area left in mobile devices for new chips. One way to solve this is to shrink the battery and leave more room for chips, but this is unacceptable since it reduces battery life. The only other possibility to continue to introduce new features and chips is to reduce the existing board area. Using IVRs, phones and tablets will have extra space on their board to include new features such as micro-projectors and ultra-accurate motion sensing. This R&D effort will have a major impact in the fast-growing smartphone and tablet market with currently one billion and 250 million in annual shipments, respectively. The industrial supply-chain to manufacture these chips in the U.S. has been either hollowed-out or will be cost-prohibitive.

---

medical diagnostics and devices, biotech, biopharmaceuticals, industrial microbiology, synthetic biology, optoelectronics, flexible electronics, quantum computing, advanced materials, and robotics. A vast array of these industries depends on a new generation of highly precise measuring equipment. More than seven hundred measurement-related barriers to technology innovation need to be addressed[179].

---

[179]An Assessment of the United States Measurement System: Addressing Measurement Barriers to Accelerate Innovation, National Institute of Standards and Technology, 2007

Many states and regions have invested in programs to create clusters in emerging industries such as optoelectronics, semiconductors, biotechnology, and medical devices. It is a model of economic development based on geographic concentrations of interconnected institutions – business, government agencies, universities – in a specific field. Social glue built around personal relationships, face-to-face contact, a sense of common interest, and "insider" status binds it together and also facilitates access to critical information. A cluster's sense of shared commitment and destiny, which transcends day-to-day business rivalries, is not easy to create. Several benefits accrue: increased productivity and efficiency by bringing together suppliers with customers, designers with engineers, and university researchers with corporate production managers, to better share information and new ideas facilitating tacit knowledge exchange. Governments should not seek to micromanage cluster creation. It is better suited to supporting and promoting these industrial networks while allowing them to develop naturally. Clusters have worked in certain places but failed in others.

Common key mechanisms for creating public or semi-public goods in industrial ecosystems are convening, coordination, pooling to reduce risk, and bridging. Inducing collaboration and spreading risks could bring a new technology to life and inject new vitality into a regional economy. Manufacturing has held in the U.S. even as so much has turned against it because firms have gained advantage from proximity to innovation and proximity to users. The gains from co-location have not disappeared.

Economic and Societal Implications

American companies continue to emerge in new technology sectors, but many keep costs down, access emerging markets and their high-skilled workers, and satisfy their investors by locating their facilities abroad, usually in Asia, instead of creating jobs here. Spending by U.S. firms on R&D outside America has grown at three times the rate of their domestic spending. Many of the nation's small and mid-size firms do not have the option to offshore research and development, and struggle to compete with foreign entities. Currently U.S. factories competitively produce about seventy-five percent of the products the nation consumes. It could meet less than forty percent of the nation's demand if the manufacturing sector remains neglected[180]. The factors driving manufacturers' choices about where to place and expand factories are: skill level and quality of factory employees especially for high-technology facilities; the presence of high-impact clusters that allow many companies to learn from one another to innovate more readily; access to nearby countries with emerging consumer markets

---

[180]Kaushal, Arvind et al – Manufacturing's Wake-Up Call, 2011

and lower-cost labor; and a reasonably competitive regulatory and tax environment.

Companies once felt an obligation to support American workers, even when it was not in their best financial interest. Profits and efficiency now trump generosity. Manufacturing executives maintain that the nation has stopped training enough people in the mid-level skills that factories need. To thrive, companies argue they need to move work where it can generate enough profits to keep paying for innovation. For technology companies, the cost of labor is minimal compared with the expense of buying parts and managing supply-chains that bring together components and services from hundreds of companies located in multiple regions of the world. Asian supply-chains have surpassed what exists in the U.S and factories in Asia can scale up and down faster. Another critical advantage is that some large emerging economies can provide engineers at a scale the U.S. cannot match.

In the last decade much of the employment in industries started in America such as solar and wind energy, batteries, semiconductor fabrication, and display technologies, is being created abroad. Between the years 1978-2007, investment in renewable energy R&D in America was $15.4 billion - after all this investment, the U.S. solar manufacturing industry employs just ten thousand people! New methods of manufacturing can increase throughput without adding jobs. Technological shifts can affect jobs in unpredictable ways - for example, because a battery-powered vehicle has far fewer moving parts than a combustion engine vehicle, fewer people are needed to design and build such a car. Battery manufacturing is radically less labor-intensive than machining lines would be.

Companies have closed major facilities in the U.S. only to reopen in China. Apple is locating production abroad but is still reaping the lion's share of profits within America. Is this going to be the American model for the future[181]? Critical strengths and capabilities that once served to bring new enterprises to life have disappeared. Are the resources that remain fertile enough to seed and sustain new growth? As scale-up of advanced technology happens abroad our capacity for initiating future rounds of innovation will be progressively enfeebled. The loss of companies that can make things will end up in loss of research that can invent them because ties that connect research in its earliest stages to production in its final phases remain vital. Many companies transitioning from venture funding to high-volume manufacturing are unable to obtain financing in the United States. Eventually they had no choice but to look for foreign investors and often moved abroad to manufacture their products. In some

---

[181]Locke, Richard and Wellhausen, Rachel (Editors) – Production in the Innovation Economy, 2013

emerging high-technology industries, it would be difficult to achieve early-stage production in the U.S. because the technical expertise, the workplace skills, equipment, and the most advanced plant layouts are no longer present here or have degraded and fallen behind the state-of-the-art elsewhere. Will separation of innovation from manufacturing

---

Boost Tower Production Throughput
Tapered Spiral Welding for Wind Turbine Towers

This publicly-funded research and development work from May 2014 to April 2017 in the amount of approximately $1.2 million to a startup based in Westminster, Colorado is to address two roadblocks to reducing the cost of wind energy: the labor-intensive construction process, and size limitations imposed by road or rail transport for turbine components. The former issue drives up manufacturing costs and reduces U.S. competitiveness against countries with inexpensive labor, while the latter forces sub-optimized tower designs and prevents turbines from growing larger and taking advantage of faster, steadier winds at higher hub-heights. This R&D effort addresses both of these problems by adapting spiral welding to wind tower production. Spiral welding is highly automated, requiring as little as 10% of the labor of the equivalent manual process. It also combines multiple operations into a single machine that can be operated on-site, eliminating transport costs and barriers. The innovation is to adapt existing spiral welders that can manufacture only straight, constant wall-thickness pipe to producing tapered, variable wall-thickness towers. A novel material geometry and automated control of machine parameters are the keys to transforming the standard system to one optimized for wind turbine tower production. With on-site spiral welding of turbine towers, significant reductions in cost of wind energy are possible.

This work, if successful, will increase the use of wind energy for U.S. electricity generation, enabled by both reduction in energy costs and increase in the number of cost-effective wind sites. Reducing the cost of tall towers enables increases in the height and size of wind turbines, allowing them to be optimized for steadier, higher speed winds. With these increases in size and optimization, decreases in cost of wind energy of 12% or more for 120-meter tall towers are possible from the current average heights of 80-90 meters. In addition, the U.S. land area for which wind energy is cost-effective can be doubled at such tower heights. Because on-site production is inherently local, manufacturing jobs are created in the communities where wind turbines are installed. Also, this method gives local production a major cost advantage over imports by producing towers that are too large to transport from port to wind farm. This allows domestic manufacturing to not only compete, but dominate in a domestic tower market worth roughly $1 billion in 2011. The startup installed the world's first tapered spiral welded tower in 2015 in Middleton, Massachusetts. Today, towers must stay below 4.3 meters in diameter in order to be shipped to the project site. Because of this constraint, towers taller than 80 meters become very difficult to design and expensive to produce. By shipping steel as flat sheets, and fabricating towers at the wind energy site, this company can produce towers in excess of 7 meters in diameter. These large diameters enable low-cost welded steel towers to reach hub-heights in excess of 140 meters.

---

allow innovation to continue at full capacity in America or will it vitiate learning and creation of capabilities that might produce future innovation here? This new global economy of fragmented research, development, production, and distribution is due to the tectonic shift in corporate ownership and control. Starting in the 1980s the dominant vertically integrated manufacturing firms began to shed their business functions such as R&D, design, detailed design, production, and after-

Miracle Materials
Shape-memory Device to Heal Muscle Atrophy

This startup based in Atlanta, Georgia was provided federal funding from July 2007 to June 2014 in the amount of approximately $1.5 million for R&D work to develop a mechanically-active soft tissue reinforcement device using a shape-memory fabric to improve the tissue quality of chronic rotator cuff tears. Shape memory polymers and alloys are a class of "smart" materials. Shape memory polymers can remember multiple shapes and transition easily between those when triggered to do so. Triggers for shape change include heat, light, and mechanical force. Shape memory alloys have a history of successful human implantation in biomedical devices such as expanding cardiac stents, guide wires and orthopedic staples. With chronic tears, the rotator cuff has degenerated to a point that prevents its ability to heal back to bone using standard repair procedures. This R&D work used the shape-memory effect to apply a continuous force on the atrophied rotator cuff tissue in an effort to promote tissue regeneration and improve overall healing capacity. The research focused on evaluating how the physical attributes of the fabric will impact the shape-memory properties and the effect of applying a force generated by the fabric's shape recovery on the quality of the rotator cuff tissue. The results of this work helped identify a fabric that can contract via shape recovery at body temperature over a time scale required for soft tissue reattachment to occur. This work has the potential to address a significant clinical problem pertaining to the treatment of chronic rotator cuff tears. The failure rate of more than 400,000 rotator cuff repairs performed each year is 20%. There are currently no reinforcement devices commercially available that can help improve tissue quality in chronic cuff tears. Thus, a reinforcement device that can apply continuous tensile force to improve tissue quality could serve as a disruptive technology and significantly impact how rotator cuff procedures are performed. This R&D effort provided the fundamental knowledge of how shape-memory materials can be applied to mechanically stimulate a biologic response in vivo, specifically the effect of tensile force on soft tissue regeneration. This shape-memory fabric technology could broadly lend itself to other clinical applications including Achilles tendon repair, hernia repair, traumatic muscle injuries, and muscular disorders. The company's vision is to develop distinctive medical devices based on a platform of "smart" materials and technology. It now has several science-driven material technology platforms to create medical devices that utilize the unique capabilities of a new generation of biomaterials to further orthopaedic surgery. These devices created from proprietary shape-memory biomaterials fusing creative design with surgical expertise can adapt inside the body and dramatically improve the technology of soft tissue fixation, fracture repair, and joint fusions. The company manufactures some of its product offerings in an in-house controlled environment using medical manufacturing equipment for custom components, assembly and packaging operations. For operations that the firm is not currently equipped to manage it seeks out contract manufacturers thus allowing it to focus on its core competencies, and reduce overall capital expenditures related to setting up manufacturing processes that it cannot support with the equipment on hand.

sales service so as to focus on their core competencies. This push came from the financial markets that derived and benefited from higher stock market valuations of leaner "asset-light" companies which had weeded out their less-profitable divisions and reduced their diversification. Among the first business functions that companies started shedding was manufacturing. This produced reductions in head-count and in capital costs that stock markets immediately rewarded. Other important contributing factors were the dismantling of border-level barriers to capital and trade flows, China's 2001 entry into the World Trade Organization, the development of Asian supply-chains of agile, dynamic subcontractors with access to huge reservoirs of cheap semi-skilled and skilled labor, new digital technology that enabled fragmentation of value-chains, and the emergence of large new Asian consumer markets requiring localization of production.

When previously large firms innovated with new products, they had the resources to scale up, but today when this happens in startups, universities and federal labs, how available is funding needed at each of the critical stages of scale-up: prototyping, pilot production, demonstration and test, early manufacturing, and full-scale commercialization? When scale-up is funded mainly through mergers and acquisitions of adolescent startups and when acquiring firms are foreign, how does the American economy benefit? The experience of some national development banks suggests that co-financing can encourage the transition of promising technologies from invention to pilot-plants for production, while yielding returns for the government. This approach is being increasingly used by other nations. A "scaling bank" can provide loans to companies investing in new domestic plants or in new technologies. State-backed financial support for new private sector manufacturing operations where conventional funding is hard to come by can often be a useful and justified part of government policy[182].

When the gains from innovation are significant but distributed thinly across many firms, it is unlikely that any single one of them will invest enough to bring it to life. Large U.S. multinational companies used to provide public goods through spillovers of research, training, and new-technology diffusion to suppliers, and pressure on state and local government to improve infrastructure. Others in the region developing new ideas and products could draw on these complementary capabilities even if they had not contributed to creating them. As these sources have dried up, market failures (large holes) in the industrial ecosystem have appeared. It is urgent to rebuild these ecosystems with new capabilities.

Future of Manufacturing in America

Many of the hundreds of thousands of small- and medium-sized U.S. manufacturing companies still operate the same way as they did fifty years ago. We need to move to high-risk, high-payoff

---

[182]Marsh, Peter – The New Industrial Revolution: Consumers, Globalization and the End of Mass Production, 2012

manufacturing of advanced products and materials[183]. However, such high-risk manufacturing requires an ecosystem of suppliers, equipment makers, and customers. New technologies could revolutionize and reinvigorate these companies, returning manufacturing jobs to America. To keep value in the U.S., the know-how responsible for creating that value needs to exist here. Once capabilities are lost to other countries, they can be hard to get back. Many manufacturing skills have a high degree of accumulated tacit knowledge that if lost is difficult if not impossible to recover.

Targeted investments to help anchor and transform manufacturing in America through shared labs, pilot plants, technology infrastructure and the creation of clusters in technology groups such as nanoscale carbon materials, next-generation optoelectronics, flexible electronics, nanotechnology-enabled medical diagnostic devices and therapeutics, advanced robotics, nano-electronics, materials by design, and bio-manufacturing could improve energy efficiency and accelerate growth[184]. Sustainability in the manufacturing sector could be the catalyst for a manufacturing renaissance in the United States.

Table 8.1 catalogs possible advanced manufacturing sectors that have the potential to create millions of high-wage manufacturing jobs in America. An array of environmental technologies, material sciences, computer-related and just-in-time production infrastructures, and the growing world of nanotechnology, have expanded product and process innovation and have so far kept American manufacturing globally competitive. Between twenty to twenty-five percent of all firms in leading-edge biotechnology and software development are manufacturers. U.S. companies are among the global leaders in developing the technical innovations that can enable countries to become more energy efficient while also achieving the economic growth so essential to improving the living standards of their citizens. America squandered its manufacturing lead in recent years in crystalline silicon wafers, liquid crystal displays, power semiconductors for solar cells, and many types of advanced batteries. Integrated photonics, lasers and modulators squeezed onto a single chip has been largely abandoned by optoelectronic manufacturers as they have moved production away from the United States[185].

Can small companies, working on some of our most promising technologies, especially at the intersection of new materials and energy, ever be able to compete with large foreign-based companies in

[183]Idelchik, Michael - How the U.S. Can Lead the Next Manufacturing Revolution, 2012

[184]PCAST Report to the President on Capturing Domestic Competitive Advantage in Advanced Manufacturing, 2012

[185]Rothman, David - Can We Build Tomorrow's Breakthroughs? MIT Technology Review, 2011

this area? If these technologies can be produced economically, they could greatly expand existing markets. The challenge then for startups

Table 8.1 Potential Advanced Manufacturing Sectors

Biotech and Industrial Microbiology	Genetically modified food/fuels/compounds; substitution of petroleum-based feedstock with environmentally friendly biodegradables
Cleantech and Alternate Energy	Electric vehicles; energy storage and conversion; rechargeable batteries; ultra-quick charging; charging station networks
Computing	Quantum computing; use of massive parallel processing
Cradle-to-Cradle Design	Materials "leased" rather than consumed
Digital Manufacturing	3D printing; use of integrated, computer-based systems; 3D visualization, analytics and collaboration tools to optimize product and manufacturing process design simultaneously
Internet of Things	Information/communication infrastructure that enables coordination of people/goods movement; do for devices/appliances/vehicles what information Internet did for computers
Machine-to-Machine Systems	Networks that allow machines to communicate with each other and use relayed information to adapt their actions to accomplish specific tasks in the face of uncertainty and variability
Manufacturing Systems	Optimized, agile, real-time production systems to respond quickly to customer needs/market changes minimizing costs and maximizing value
Medical Devices	Providing physicians the best tools to diagnose/treat patients; nanoscale-enabled medical diagnostic devices/ therapeutics
Nanotechnology and Mechatronics	Integration of mechanical systems, smart materials, and electronics; flexible-, nano-, and organic-electronics
New Materials Manufacturing	Bio-manufacturing; manufacturing techniques to turn new materials into products for use in energy, consumer electronics, computation and biotech; advanced materials, nano-scale carbon materials, and materials by design
Optoelectronics	Next-generation, energy-efficient lighting; LED manufacturing for display, lighting and sensing
Pharmaceuticals	Accelerate manufacturing of drugs and vaccines through rapid, flexible and cost-effective manufacturing systems
Robotics	Devices that are partly/largely autonomous, that interact physically with people or their environment, and that are capable of modifying their behavior based on sensor data
Synthetic Biology	Designing and constructing biological devices and systems for useful purposes
Wireless Devices	Integrated microsystems capable of wirelessly measuring or controlling physical parameters, interpreting data, and communicating information

is to figure out a way to produce their technologies using current manufacturing know-how while developing products that are radical enough to disrupt established technologies. Still, such startups face the daunting truth that scaling up innovations into manufacturing operations can take hundreds of millions of dollars. Economies based primarily on services will be second-tier, and services not built largely on the back of a vibrant manufacturing sector, and the breadth and depth of the

---

Additive Manufacturing
Facilitate 3D Printing of Industrial Parts

This taxpayer-funded research and development work from April 2015 to March 2017 in the amount of approximately $730,000 to a startup based in Dallas, Texas is to understand interface chemistry and adhesion phenomena in a special class of low-viscosity resins to produce a range of mechanically tough materials that are 3D printable via stereolithography. A significant problem with current stereolithography approaches is that successive printed layers do not sufficiently adhere together, leading to large reductions in toughness in soft, viscoelastic and stiff materials. This R&D effort explores the tradeoffs between molecular architecture, reactivity, resin viscosity, and key printing parameters to develop improved materials to enable tougher printed parts than industry standards, along multiple axes of deformation at comparable current industry printing speeds, with feature sizes well below a hundred microns. This work is expected to further enhance the thermomechanical properties of the company's polymers by incorporating proper additives into its processes to control color, shelf life, and aesthetics. The impact of this work will be to facilitate the making of tough, 3D printed parts that can be directly manufactured through additive processes that are already commercially available. Additive manufacturing has the potential to revolutionize the way parts are produced by streamlining product design, production, and validation leading to low production costs and accelerated lead times. The penetration of additive technology into industrial processes has been greatly slowed by the current inability to 3D-printed components of any stiffness with material properties on par with traditionally manufactured parts. Particularly, 3D printed materials tend to tear or fracture more readily between successive printed layers. The materials and processing techniques thus developed by this startup will help drive additive manufacturing into large volume, yet customizable market sectors to increase efficiency and productivity across industries. The printed parts resulting from this company's printable materials will further U.S. manufacturing by lowering production costs, increasing product performance and reshoring advanced manufacturing.

---

ecosystem that grows from manufacturing will not lift or sustain a nation's economy. To compete and succeed in the global economy, American manufacturers should be the world's leading innovators, and the U.S. should be the best place in the world to manufacture and attract foreign direct investment. We should expand access to global markets to enable U.S. manufacturers to reach the ninety-five percent of consumers who live outside our borders. Manufacturing in America should have the workforce that the twenty-first century economy requires by promoting a system of nationally portable, industry-recognized skills credentials[186].

---

[186]A Growth Agenda: Four Goals for a Manufacturing Resurgence in America, National Association of Manufacturers, 2012

One challenge will be to develop the tools and techniques that would enable large-scale manufacturing equipment and materials to efficiently produce low-volume lots, down to units of one. Prototyping new products in exactly the same high-volume manufacturing facility

---

Graphene, Emerging Super Material
Roll-to-roll Production of Uniform Graphene Films

This publicly-funded research and development work from September 2013 to February 2016 in the amount of approximately $902,000 to a startup based in Philadelphia, Pennsylvania is to develop technology for the roll-to-roll production of continuous graphene films. This graphene production technology is based upon innovations in graphene synthesis and handling that addresses critical deficiencies that has so far limited industrial manufacture of graphene. The synthesis process is performed at atmospheric pressure, allowing roll-to-roll graphene formation on continuous tapes of copper foil passed through the growth region. This eliminates the need for an expensive vacuum furnace and allows fabrication of graphene films larger than the furnace size. The graphene handling process developed during earlier-stage research funding enabled the transfer of graphene sheets from the metal catalyst to nearly any smooth surface without high-temperature steps, and without the use of harsh chemicals. Most importantly, the graphene transfer preserves the original metal substrate for reuse. The reusable substrate dramatically reduces the cost of graphene production and eliminates the largest source of waste in the process. In this R&D effort, continuous film processes for graphene synthesis and transfer to new surfaces, and the design of a large area roll-to-roll graphene production system will be demonstrated. This work, if successful, will result in industrial scale availability of high-quality, low-cost graphene sheets. Graphene has been called "the miracle material of the 21st century" because of its light weight, extraordinary strength and superior conductivity. Transparent, electrically and thermally conductive, flexible, and gas impermeable, graphene has innumerable proposed applications such as flexible transparent conductors for displays and photovoltaics; high frequency electronics for communications; chemical and biological sensors; as a corrosion barrier; for filtration and water desalination; and energy storage. Industrial quantities of graphene films will enable the development of these and other applications with the advantages of cost, quality, and design flexibility over competing concepts. This startup's proprietary chemical vapor deposition process works at atmospheric pressure compared to all other competing techniques that require a more expensive process utilizing a vacuum environment. The company is uniquely positioned to mass produce graphene on a commercial scale and at an economically viable cost. Its innovative, patent pending, manufacturing process is also able to transfer graphene to nearly any substrate. The startup's mass production capabilities will facilitate market disruption within the biosensor, desalinization, and electronics industries. Its first graphene-based product - grids for transmission electron microscopy - is already on sale to a development partner.

---

using the same materials and processes that will be used in the final product could revolutionize production speed and efficiency. It could eliminate the time, cost, and risk that arises from transitions between stages that include prototyping, early production runs, limited- and then finally large-scale manufacturing. These transitions currently require extensive rework, and are the source of production delays, surprises, and cost overruns. Design tools that dramatically improve the existing systems engineering, integration, and testing process for complex electromechanical, cyber-physical systems that represent the bulk of manufactured products today will reduce the need for expensive build-test-design cycles. Manufacturing facilities similar to today's semiconductor foundries can lead to rapid reconfiguration to accommodate a wide range of design variation substantially reducing the time required to go from design to product leading to flexible, programmable, and potentially distributed production capabilities able to accommodate a diverse range of systems and system variants, rather than requiring separate facilities for single products. In similar fashion as open-source code development, open-source collaboration environments will democratize the design innovation process by engaging a vastly larger pool of talent than current industry models.

We are fond of looking to the future, because our secret wishes make us apt to turn in our favor the uncertainties which move about in it hither and thither

– Goethe

Researchers are making headway on artificial intelligence and society stands on the cusp of unprecedented change, with great advances in biology, life sciences, healthcare, materials, robotics, machine learning and perception, powering systems that rival or exceed human capabilities. Driverless cars, robotic helpers, and intelligent agents that promote our interests have the potential to usher in a new age of affluence and leisure but the transition may be protracted and brutal unless we address the two great scourges of the modern developed world: volatile labor markets and income inequality. Innovative, free-market adjustments to our economic system and social policies can avoid an extended period of turmoil[187].

What are the jobs of the future? How many will there be? And who will have them? One might imagine and hope that today's industrial revolution will unfold like the last: even as some jobs are eliminated, more will be created to deal with the new innovations of a new era. Unfortunately, as technology continues to accelerate and machines begin taking care of themselves, fewer people will be necessary. Artificial intelligence is already well on its way to making "good jobs" obsolete: many paralegals, journalists, office workers, and even computer programmers are poised to be replaced by robots and smart software. As progress continues, blue and white collar jobs alike

---

[187]Kaplan, Jerry – Humans Need Not Apply: A Guide to Wealth and Work in the Age of Artificial Intelligence, 2015

will evaporate, squeezing working- and middle-class families ever further. At the same time, households are under assault from exploding costs, especially in education and healthcare that have not yet been transformed by information technology. The result could well be massive unemployment and inequality as well as the implosion of the consumer economy itself. What happens to the consumer economy when you take away the consumers? Past solutions to technological disruption, especially more training and education, are not going to work, and we must decide, now, whether the future will see broad-based prosperity or catastrophic levels of inequality and economic insecurity[188].

Grand challenges can serve to focus collaboration between the private and public sectors; help tackle important problems related to the environment, energy, health, and education; and help create jobs of the future. Table 9.1 lists global challenges based on an extension of the U.S. National Academy of Engineering's list of grand challenges[189] - solutions to these can lead to whole new industries. In this context, there is the brain research through advancing innovative neuro-technologies (BRAIN) initiative to revolutionize our understanding of the human mind and uncover new ways to treat, prevent, and cure brain disorders like Alzheimer's, schizophrenia, autism, epilepsy, and traumatic brain injury; the SunShot Grand Challenge, to make solar energy cost-competitive with coal by the end of the decade; the Electric Vehicle Everywhere Grand Challenge, to make electric vehicles that are as affordable as today's gasoline-powered vehicles within the next ten years; the Asteroid Grand Challenge, to find all asteroid threats to human populations and know what to do about them; and the Grand Challenges for Development[190].

Nevertheless, in the near-term, what all Americans want is a good job - we need to create twenty-one million new jobs by 2020 to return to full employment. This can be accomplished by further building our innovation capacity[191] focusing on technologies that are strategic and not too late in terms of technology life-cycle transitions, and that address grand challenges such as those presented in Table 9.1. An example is offered by optoelectronics, an increasingly important industry because of the forthcoming migration of computers to photonics-based technologies that is in the process of transitioning

---

[188]Ford, Martin – Rise of the Robots: Technology and the Threat of a Jobless Future, 2015

[189]National Academy of Engineering of the National Academy – Grand Challenges for Engineering, 2008

[190]https://www.whitehouse.gov/administration/eop/ostp/grand-challenges

[191]Atkinson, Robert and Andes, Scott - The Atlantic Century II: Benchmarking EU and U.S. Innovation and Competitiveness, Information Technology and Innovation Foundation, 2011

from a discrete to an integrated technology format. Monolithic integration has performance and cost advantages and could potentially be a growth industry for the United States. Meanwhile, the projected job-creating sectors in the 2012-25 timeframe are healthcare; business

Table 9.1   Global Grand Challenges

Access to Electricity for All
Advance Health Informatics
Advance Personalized Learning
Affordable Housing For All
Banish Hunger and Poverty
Develop Carbon Sequestration Methods
Engineer Better Medicines
Engineer Advanced Tools for Future Scientific Discovery
Efficiently Fight Pandemics
Enhance Virtual Reality
Good Health and Nutrition For All
Have All the World's Children Reading
Make All Voices Count
Make Solar Energy Economical
Manage the Nitrogen Cycle
Mitigate Climate Change Effects and Reverse Environmental Degradation
Power Agriculture
Prevent Nuclear Terror
Provide Energy from Fusion
Restore and Improve Urban Infrastructure
Reverse-engineer the Brain
Save Lives at Birth
Secure Cyberspace
Secure Water for Food, Drinking, Hygiene, and Industry

services (includes high-tech services, science and technology, and financials); leisure and hospitality; manufacturing; retail; and infrastructure/construction[192].

Evolution of Discovery-based Innovation

The innovator-entrepreneur remains an American icon and is venerated. The United States invented and applied all sorts of crucial technologies, as displayed in Table 9.2, showing that many of the twentieth century's greatest engineering achievements were first developed and commercially deployed in America. Although a number of countries around the world are progressive in their adoption of technology, the U.S. still sets the pace for technological innovation.

---

[192]McKinsey Global Institute – An Economy That Works: Job Creation and America's Future, 2011

This can be attributed to the same risk-taking, entrepreneurial drive on which this nation was built. Program officers in the federal government together fund more high-technology startups in a year than a venture capitalist would in a lifetime. This allows them, among other things, to spot trends in early-stage investing. One particular development raises

Table 9.2   Greatest 20th Century Engineering Achievements

Air Conditioning and Refrigeration
Airplane
Automobile
Computer
Electrification
Electronics
Health Technology
Highways
Household Appliances
Imaging
Internet
Laser and Fiber Optics
Materials Science
Mechanized Agriculture
Nuclear Technologies
Petroleum and Petrochemical Technology
Radio and Television
Spacecraft
Telephone
Water Supply and Distribution

concern in that foreign multinationals peek into our publicly-funded small business portfolio, are intrigued by various pieces of technology we fund, perform diligence on companies of interest, and quickly make decisions to invest millions, in some cases tens of millions, in our extreme technical risk early-stage technology startups. This facilitates the incorporation of American taxpayer-funded technology into their production lines, a reflection of the "discover here, make there" phenomena. To counteract this dire trend requires laser-like focus, and a comprehensive, coordinated, integrated, and sustained approach, not dithering while other nations make definitive progress.

Evolution of the federally-funded startups portfolio with possible focus areas suggested by Table 9.3 can be better guided by leveraging investments in basic science and technology research made in the over one hundred federally-funded research centers, by better integration of the various already existing industry-university partnerships, and by working more closely with university research parks, incubators, and accelerators. Additionally, charts similar to Table 7.2 can be developed for the suggested technology sectors for a more strategic approach, and the sourcing of innovation around the country, especially in metropolitan areas, can be better targeted. Helping to close critical gaps in hollowed-out supply-chains and scaling for production in America are most worthy of consideration. A focused, targeted exploration of new markets worldwide for example by roll-up and clustering of startups by technologies will be beneficial. A more integrated approach to customer discovery, business modeling, and entrepreneurial education will increase proposal quality that would also better leverage prior federal basic research investments.

Table 9.3   Federally-funded Startups Portfolio Evolution

3D printing
Advanced health informatics
Advanced manufacturing
Advanced materials for energy, medicine, food, sensing, computing, housing, clothing
Autonomous vehicles
Biomimetics
Biotechnology, synthetic biology, and industrial microbiology
Clean technology – nuclear fusion; electric transportation; wind; solar; geothermal; hydrogen; hydroelectric; hydrothermal
Climate change mitigation - geoengineering, carbon sequestration; air and water pollution reduction; clean water and desalination
Data science - big data, small data, open data, data exhausts, data mining, data brokerage
Food technology – new foods, new agriculture, space-grown foods
Genomics; personalized medicine for humans, animals, and plants; stem-cell based cures
High-speed computing; quantum computing
Human-computer interfaces – artificial intelligence, artificial reality, virtual/augmented reality
Internet of things; cybersecurity
Medical devices; drug discovery and delivery
Nanotechnology; nanoscale measurement and instrumentation
Next-generation semiconductors – flexible, organic, biologic
Optoelectronics, integrated photonics, solid-state lighting
Personalized learning; massively-scaled education
Regenerative medicine (tissue engineering); organ printing
Robotics; drones
Wireless technology; wireless charging; wireless electricity

The discovery-to-innovation-to-market pathway as exemplified by the assistance that can be provided to startups and documented in this volume can serve as a national model of an innovative public-private partnership for enhancing the growth of early-stage companies[193]. One ambitious goal can be to scale and tailor this effort

---

[193]Note: The underlying theme for the six examples presented in this chapter is to highlight the potential of technologies being developed by taxpayer-funded startups to seed whole new industries in America.

across multiple state-level programs, in order to help seed a million high-wage, high-skill jobs every year. To accomplish this, an important

---

Modernize America's Electricity Grid
Solid-state Transformer as Energy Router

This publicly-funded research and development work from September 2014 to February 2017 in the amount of approximately $1.1 million to a startup based in Rayleigh, North Carolina is to combine various research ideas in order that it may commercialize the company's breakthrough product for electric utilities, the grid energy router. This product will be the cornerstone for an evolved grid that can integrate renewable energy and storage, offer dynamic efficiency gains, and intelligently route power. Early work in power electronics by this company only focused on the area of voltage conversion limiting commercialization and manufacturability. This R&D effort will combine three years of market research, the voice of the energy utilities customer, a unique product roadmap, and cutting-edge research in the areas of feature implementation and voltage conversion. Electric utility requirements have been incorporated: highly efficient, cost-competitive, manufacturable within a specific market window, and scalable both to high-power and high-voltage. This work facilitates a cost-effective and electrically-efficient product design ready for industrialization and grid integration, while simultaneously incorporating valuable features that justify utility expenditure and market needs. This effort, if successful, will result in the development of a cornerstone for reliable electricity and a modernized grid able to evolve alongside emerging customer demands. Reliable electricity is a key component of an industrialized market, critical for the information age, and an enabler for the development of non-industrialized regions. This R&D effort will enable better scientific understanding and technology development related to energy storage, savings, and efficiency; photovoltaic and other renewable energy generation; and electric vehicles and their correlated fast chargers. We need advanced power delivery that is essential for safe and reliable electricity. Our electrical grid is outdated. It was designed for a different time and a different consumer. This company was formed in order to develop, design, and deploy breakthrough solutions that will modernize power systems and enable them to evolve continuously. The startup's mission is to revitalize power delivery systems for critical applications by combining breakthrough technology that enables distributed resources, enhances reliability and resilience, and improves efficiency. Mindful consumers are increasingly engaged with their electricity use, monitoring cost and efficiency, as well as environmental effects. In today's electric grid, the solution is to insert a rich set of features at the distribution transformer downstream of the substation, introducing granular power control at key intersections, and combining innovative power electronics-based devices with proprietary local software and communications capabilities.

---

first step is to form a powerful and illustrious Board that will evangelize, persuade, and convince decision-makers at the highest levels of the financial sector, corporate America, universities, and federal, state, and metropolitan governments, of how critically important it is to assist federally-funded startups more efficiently take their products and services to market. In this near-term evolution, federal program officials can further develop their efforts in the following areas:

Strategic Industry Partnering

To provide comprehensive coverage of the startup portfolio through the use of powerful technology-specific networks that facilitate business intelligence, industry relationships, and effective communications requires partnering with industry in a methodical, strategic fashion. The small business innovation research program must continue to expand and automate the build-out of networks with emphasis on an evolving set of technology segments such as those listed in Table 9.3. Its efforts must align with corporate innovation models; it must develop supply-chain and distribution channel maps in these technologies to assist startups with partners, customers, joint development agreements, acquisition of corporate intellectual property and donated patents; and it must facilitate rapid growth via effective use of an innovation web portal. This will require in-depth, always ongoing portfolio analysis using the most appropriate tools for post-monitoring of public investments.

The program can identify clusters, establish profiles, and perform technology landscape mapping, to reveal opportunities for startups at the intersection of emerging technologies and human needs. It can develop business intelligence by partnering with market research and consulting firms. It must perform analyses to enable more targeted approaches to industry partnerships - industry outreach has been mostly a "machine-gun" approach, and overly focused on large multinational companies. The companies that have in fact done deals with federally-funded startups have not necessarily been such corporations. For a good part of this portfolio, our startups sell into less visible parts of certain supply-chains; knowledge of these supply-chains would be helpful to startups and for industry outreach activities. This activity can also be "crowd-sourced" from our startups using a web-based innovation platform. This fully automated web platform will serve as a startup marketing vehicle and lead to synergistic clustering and/or roll-up opportunities all assimilated in a single database with google-like search capabilities and other social media features. The program can continue to conduct technology-specific events such as startup-industry centered roundtables and workshops, and facilitate our small businesses to periodically participate in trade shows and professional society meetings.

Innovation Research Consortia

Technology licensing offices (TLO) are responsible for commercializing university research and development, some of which is funded by federal agencies. These entities are uniquely capable of identifying promising startups around federally supported research. They are often knowledgeable about the R&D strengths of their university, and the most promising entrepreneurial activity in those areas. The small business innovation research program can target technology licensing offices to help increase the number and quality of proposals, and find ways to better leverage disclosures from federally-funded research. The focus will be on supporting and/or partnering with startups. University research, researchers, and university-generated intellectual property can be valuable to startups, and it can help them become more competitive.

Biomimetics for New Gripping Materials
Manufacturing of Bio-inspired Polymer Fiber Arrays

This taxpayer-funded R&D effort from April 2012 to June 2016 in the amount of approximately $669,000 to a startup based in Pittsburgh, Pennsylvania is to develop pilot-scale production processes and systems to enable large-scale fabrication of continuous arrays of elastomeric micro- and nano-scale fibers with complex geometries. Inspired by hairs that occur naturally on gecko feet, these elastomeric fibers demonstrate strong adhesive, shear, and peel strengths over a wide range of test substrates. Unlike other classes of adhesives such as pressure-sensitive tapes, these biologically-inspired adhesives can be repeatedly used over thousands of test cycles with little contamination and performance degradation over the material's lifespan. This class of material has until now only been produced through expensive fabrication processes such as photolithography, chemical etching, or time-consuming batch molding processes. In this R&D work, a roller-based molding and peeling process for high-speed, continuous, and large-area manufacturing of high aspect-ratio and three-dimensional micro- and nano-scale fibers with a compliant backing layer will be developed using elastomer materials.

This effort, if successful, will provide a low-cost, high-volume process to mass produce continuous arrays of elastomeric fibers with complex geometries for applications in apparel, sporting equipment, healthcare, defense, industrial clamping, and consumer goods. These fibers will provide strong reversible adhesive or enhanced shear interfaces that are resistant to contamination and maintain their adhesive ability over the product lifespan. The company's product is the world's first dry adhesive whose power derives from its advanced, nano-level geometric design made up of millions of microscopic structures which form molecular-level bonds with a mating surface. When multiplied across millions of fibers, this adhesive forms a strong, dry, residue-free, and repeatable connection solution. Unlike adhesives which create chemical bonds and are sticky or permanent, and unlike vacuum or suction cup approaches that are costly or temporary, this offers a dry solution that lasts, yet provides repeatable performance. Since the geometric design can be implemented in a variety of materials and mated to a variety of surface types, this product can cut across multiple product categories, from adhesives and gripping material to fasteners. Since it works at the microscopic level, it creates low profile, conformable connections which can be hermetic. Encouraged by the results of nearly a decade of academic research in understanding and synthesizing biologically-inspired adhesives, a university professor founded and spun out this startup from Carnegie Mellon University in 2009.

Certain web-based networks have members that include technology licensing offices from dozens of U.S. universities. To the extent that there are startups in technology areas of a particular universities' strength, these offices could support that startup with intellectual property or industry contacts. This is a "win-win", with these offices leveraging startups to commercialize their intellectual property, and with startups leveraging the TLO's relationships with university researchers and industry. A deliberate long-term campaign can be instituted to inform TLOs of potential small business proposals. These could incent these entities to promote startups or promote related IP to federally-funded startups that could leverage it. Access to entrepreneurs is a top priority at some technology transfer offices. The small business-technology licensing office connection would be valuable in this regard to facilitate IP exchange, business modeling, and outreach to further develop entrepreneurial networks. These can be manifested through a web-based platform. In addition, the small business program can partner with university-based incubators such as the Stanford University-associated StartX and the Advanced Technology Development Center at Georgia Tech, to hold competitions, offer prizes, and conduct regional events such as case study-based sponsored activities and small business-centered workshops. Prize competitions such as the Robobowl at Carnegie-Mellon University can be expanded to other technology areas, and the pitch competition for would-be entrepreneurs at the seventeen Engineering Research Centers listed in Table 3.6 can be scaled to over hundred other federally-funded research centers.

Incubators for Manufacturing Innovation

To make what you invented, the small business innovation research program can create manufacturing "incubators" to develop and execute the lab-to-manufacture model which in essence is "born here, make here, and sell everywhere". We must retain the ability to make things. It is about jobs, and it is about finding a new path forward to monetize American innovation by utilizing existing assets, leveraging the current workforce and intellectual skills, preserving capital, reducing risk, and improving predictability. This will help anchor subsequent manufacturing in the United States through shared labs, pilot plants, technology, infrastructure, and the creation of clusters. The program can make more broadly available the unique research, innovation, and skilled workforce capabilities that exist, for example, at the Eastman Business Park. This facility can serve as an infrastructure for a manufacturing "incubator" to ensure that the next generation of processes and products will not only be invented in the U.S., but scaled up and manufactured in America as well. It will allow startups to develop their businesses from the lab-scale prototype stages of innovation to the later stages of production and commercialization.

The President's Council of Advisors on Science and Technology recommended that such efforts can "be expanded nationwide to support startups emerging from Federal advanced manufacturing research programs. Support could include expanded mentoring initiatives, efforts to mobilize corporate partners interested in supporting beta testing of products, and/or the creation of a seed fund for advanced manufacturing start-ups. A focus on creating a national network to foster stronger start-up interactions with leading manufacturers would be a key element of a network that would complement and aid the efforts of individual universities to integrate a manufacturing focus into innovation ecosystem programs to create and support start-ups[194]." Such an expansion envisions the establishment of a vertically integrated manufacturing innovation ecosystem consisting of three elements as described below:

i.   Discover, Map, and Help Repurpose Under-Utilized Manufacturing Assets for Startups

For a variety of reasons, a number of mid-size and large corporations have manufacturing assets, within a once vibrant local innovation ecosystem, that are at risk of extinction unless a concerted effort is made to sustain or retrofit them. At the same time, a sizable number of manufacturing oriented small businesses in this portfolio of federally-funded startups need access to such facilities to develop and demonstrate that their production can be scaled for domestic manufacture. While some such sites are well known – such as the coating assets of Kodak in Rochester, New York; the Whirlpool/Maytag facility in Newton, Iowa now converted to a wind-turbine manufacturing facility; North Carolina's manufacturing economic plan to repurpose from previous dependence on textile and furniture industries; under-utilized aerospace industry manufacturing facilities in California and Florida; the semiconductor manufacturing facilities in Albany, New York – there are many others that exist nationwide but are less familiar.

We should conduct a thorough mapping of under-used manufacturing facilities around the U.S., with special focus on small- to medium-size manufacturing companies, develop a national database that identifies such entities (profiles, key contact information) that can serve as manufacturing assurance nodes, develop affordable access mechanisms, and significantly increase the speed of commercialization for startups. Many startups could benefit from this, and because they are unaware of such facilities, they consequently have higher funding requirements, and often times the production occurs outside America.

The program can set up mechanisms such as prizes and competitions for startups that use such equipment, much of which is dormant or considered a liability because the company has no use for it anymore, and most likely will be sold for scrap. Better then to leverage existing manufacturing equipment and furloughed skilled workers to create new products and retain high-wage jobs. There are significant manufacturing assets at these companies that are open for "toll manufacturing" for the first time. Additionally, skilled but laid-off manufacturing workers could partner with such startups and be a source of manufacturing innovation. Creating relationships with entrepreneurs, especially those in the open manufacturing and open hardware movements, and makerspaces, can lead to high-quality

---

[194]PCAST Report to the President on Capturing Domestic Competitive Advantage in Advanced Manufacturing, 2012

advanced manufacturing proposals. Besides helping startups mitigate risks, this effort will assist them to re-shore, to scale-to-manufacture in America, to successfully overcome, in the case of manufacturing, the two valleys-of-death hurdles (explained below), and thus provide a robust, sustainable high-wage job-multiplier.

ii.   Establish "Small-to-Small" Advanced Manufacturing Cooperatives

The commercialization model of small businesses supplying fresh technologies to large corporations may not be as effective in the current hyper-connected world where the only competitive advantages that matter may be "ideas" and "speed-to-market". Furthermore, there is the issue of small family-owned manufacturing shops in America, an approximate analog to the Mittelstand companies in Germany. The trend seems to be that as baby boomers retire, family lineage of such manufacturing firms is tending to end. We must sustain or retrofit their micro-ecosystems and supply-chains in cases where these businesses are not transferred to entities that are willing to continue to operate them. This presents an opportunity for entrepreneurs and a new generation of owners and managers. Our startups along with such family-owned manufacturing concerns and other U.S.-based small businesses may be able to grow and compete far more effectively in this dynamic environment if they can cooperate and collaborate amongst themselves. This would involve formation of advanced manufacturing cooperatives to generate and sustain competitive manufacturing advantage via dynamic pooling of resources, merger and acquisition activities among startup members, and shared approaches to infrastructure, intellectual property, contract research, and performance metrics.

iii.   Assist Grantees to Leverage the National Network for Manufacturing Innovation (NNMI)

A network of institutes focused on reducing and solving pre-competitive industrial problems, the vision is to bring together institutes of manufacturing excellence where our most advanced engineering schools and our most innovative small businesses and manufacturers collaborate on new ideas, technologies, methods, and processes. The federal investment in NNMI serves to create an effective manufacturing research infrastructure for U.S. industry and academia to solve industry-relevant problems. Institutes will draw together the best talents and capabilities from all partners to build proving grounds where innovations flourish and American domestic manufacturing advanced.

NNMI consists of linked Institutes for Manufacturing Innovation (IMI) with common goals but unique manufacturing technology concentrations. In these institutes, industry and startups, academia, and government partners leverage existing resources, collaborate, and co-invest to nurture manufacturing innovation and accelerate commercialization. As sustainable manufacturing innovation hubs, IMIs will create, showcase, and deploy new capabilities, new products, and new processes that can impact commercial production. NNMI will allow new manufacturing processes and technologies to progress more smoothly from basic research to implementation in production. IMIs will offer facilities comprising an "industrial commons" - the R&D, engineering, and manufacturing capabilities needed to turn inventions into competitive, commercial products - and accelerate the formation and growth of small- and medium-sized enterprises. IMIs will provide shared-use facilities with the goal of scaling up laboratory demonstrations and maturing technologies for manufacture.

The focus is on scale-up, effectively tackling the missing middle, the scale-up gap, the second valley of death, to engage small and medium manufacturing enterprises, to enable technology, manufacturing processes, and the use of advanced materials in different industry sectors such as medical devices, robotics, optoelectronics, automotive, aerospace manufacturing, and industrial microbiology. It will help address relevant manufacturing challenges on a production-level scale, help startups develop innovative methodologies and practices for supply-chain integration, and reduce the cost and risk of commercializing transformative new technologies.

An example of a manufacturing "support network" is provided by Maker's Row, an online directory that contains profiles of some eight thousand U.S.-based manufacturers[195]. On Maker's Row, manufacturers get a basic listing for free or pay about fifty dollars a month for more elaborate profile pages that may include photos of finished products and video tours of factories. Two hundred dollars a month buys priority placement in search results. Registered users – the sight has eighty thousand – get basic manufacturer information for free or pay twenty dollars a month, for access to a detailed search feature, customer reviews, and direct factory contacts.

One of the co-founders started his own watch brand in 2007 - his small business had little leverage with the huge Asian factories filling his orders; for startups, there is no red carpet overseas. He switched to leather goods but struggled to find domestic factories to handle growing demand for his company's wallets, belts, and accessories. He came up with an idea for a site that would help connect designers with American manufacturers; many of the site's users are 9-to-5ers with ideas for products but no roadmap for how to turn them into a business. The registration-and-fee structure helps weed out the frivolous; to guide more committed newbies, Maker's Row also offers educational content, including a free six-week online course called Production 101. This 25-person company has opened an

---

[195]Business Week, November 2015 – Maker's Row, Help for Small Business Manufacturing Needs

online store showcasing the wares of its designers; it is also piloting a payment system that factories could use for billing. Maker's Row is helping to create an ecosystem that is lowering barriers for small and medium-size businesses. Many American manufacturers want to diversify their client base and have learned that means getting online. Some small manufacturers talk about how the recession underscored the risk of relying on a small number of big clients. Firms are using Maker's Row to get their name out, like a form of advertising. This Made in America emphasis has turned Maker's Row into a darling of politicians.

Startup Exports

More than seventy percent of the world's purchasing power is now located outside the United States. Increased foreign demand for U.S. goods translates into more jobs, greater revenue, and better wages for American businesses and workers. Tradable sectors of the economy drive wealth, boost productivity, and grow local industries but have severely underperformed in the last few decades. To reverse this trend, states and metropolitan areas are helping firms in key industry sectors grow and innovate by selling their goods and services globally. Many leaders in states, cities, and metro areas across the country are exploring ways to help their firms tap into expanding markets worldwide to grow jobs at home by developing effective action-oriented metropolitan export plans and initiatives customized to their region's unique assets and capacities.

Exports accounted for nearly half of the nation's economic growth in the first year of the post-recession recovery and can power economic growth over the long haul. In 2013, total exports made up only 13.5% of our GDP, compared with 50.7% in Germany, 30.1% in Canada, 26.4% in China, 24.8% in India, and 14.7% in Japan. Less than one percent of American companies export, and of those, just over half export to more than one country[196]. Export income supports the purchase of local goods and services, exporting firms pay 9.1% more than jobs in firms that export less, and in the U.S. a dollar of exports produces twice as much employment as a dollar of domestic consumption. One study finds that every $1 billion in new exports creates 5,400 additional jobs.

Exporting spurs innovation; small- and mid-sized firms that export tend to innovate more in products and processes than non-exporters. Table 9.4 provides 2013 data on the top ten U.S. exports. At the Export-Import Bank, financing authorizations for exports grew 34% to $32 billion in 2011, with 85% of transactions to small- and medium-sized enterprises (SMEs). In America, a record 287,000 SMEs, 98% of all exporters, exported in 2010, a total increase of more than 16,600

---

[196]Katz, Bruce and Bradley, Jennifer - The Metropolitan Revolution: How Cities and Metros are Fixing Our Broken Politics and Fragile Economy, 2013

SMEs over 2009. Moreover, SMEs continue to grow their share of overall U.S. exports, composing 34% of total export value in 2010, up from 27% in 2002. This trend is affected by the prominent role SMEs

Table 9.4   Top U.S. Exports

Sector	Export Value ($B)	% of Total Exports
Machinery	213.1	13.5
Electronic Equipment	165.6	10.5
Mineral Fluids including Oil	148.4	9.4
Vehicles excluding Trains and Streetcars	133.6	8.5
Aircraft, Spacecraft	115.4	7.3
Optical, Technical, and Medical Apparatus	84.3	5.3
Pearls, Precious Stones/Metals, Coins	72.8	4.6
Plastics	60.8	3.9
Organic Chemicals	46.5	2.9
Pharmaceutical Products	39.7	2.5
	1080.2	**67.4**

play in competitive industries, such as environmental technologies and medical equipment. The United States must re-double efforts to expand its national base of SME exporters to sustain this historic progress. SME exporters outperform their non-exporting SME counterparts in several measures including higher revenues, faster revenue growth, and higher labor productivity.

Large corporations are generally more capable of accessing global markets, but startups typically do not have the resources to explore such options. A program aligned with the 2010 National Export Initiative can help connect U.S.-based small businesses to overseas markets. Federally-funded startups can leverage this initiative to create mechanisms that build cooperative and collaborative ties among identified global entities and these startups can then begin or expand exporting. It will increase our startups' awareness of export opportunities, export finance programs, and other federal, state and metropolitan resources.

Private Equity Syndicates

An analysis of recent investments by the top fifteen venture capital firms (Table 5.3), the trend of negative returns-on-investment, decreasing capital under management, increasing appetite for ever-decreasing risk profiles, and vanishing exit options suggests a broken venture capital model, especially in certain critical, strategic high-technology sectors. There is an urgent need for creative new financial instruments such as equity crowd-funding, new types of bonds, and prize mechanisms to serve as technology-specific investment vehicles to augment capital in technology areas that venture capital firms are unwilling to fund. Federal programs that fund startups can mitigate these trends by:

a. Establishing formal partnerships with the Angel Capital Association and the National Venture Capital Association to leverage their existing investor networks; making robust and wide-ranging relationships with private equity firms and investment banks; and looking for opportunities to syndicate with corporate venture capital funds.

b. Helping establish ten dedicated $100 million each technology venture funds in ten technology sectors that have the potential to seed and create whole new industries of the future such as robotics, industrial microbiology, optoelectronics, advanced semiconductors, and medical devices.

c. Working with state and metropolitan entities to help create new financial instruments such as social impact bonds, job creation bonds, innovation bonds, and prize money mechanisms to nurture local high-tech startups; establishing formal relationships with existing crowd-funding platforms; and perhaps in time creating federal and state crowd-funding platforms.

d. Creating a parallel private startup program to help fund nascent companies that angels and venture capital firms are unwilling to invest in due to excessive technical risk; these same startups can also be encouraged to apply to already existing federal programs; and forwarding proposals that are considered too risky for venture capital firms and angels in terms of technical feasibility to such federal programs for possible consideration/investment.

Regional Technology Clusters

To gather and leverage the benefits of efficiencies of scale, federal startup programs can, when appropriate, align with state and metro technology innovation initiatives. To establish a nationwide footprint, these programs can systematically explore and work with local supply-chains and support structures to ignite regional technology clusters across America. Table 9.5 provides a sample listing of such clusters. Additionally, such programs can identify and leverage state technology funds and legal mechanisms to foster technology innovation and promote local entrepreneurship.

Barrier to Innovation

The time is overdue to fix the patent system. Governments protect patents because they are held to promote innovation but there is plenty of evidence that they do not. In agriculture and in other

Table 9.5　　　　Technology Cluster Examples

Akron, OH	Rubber
Albany, NY	Semiconductors
Baltimore, MD	Health Sciences
Bethesda, MD	Health Sciences, Biomedical, Medical Devices
Boston, MA	Robotics, Biotech, Nanotechnology
Cleveland, OH	Medical Devices, Water Tech, Energy Services
Dallas-Fort Worth, TX	Semiconductors, Wind Energy
Denver, CO	IT, Telecom, Life Sciences, Optics, Aerospace
Detroit, MI	Automotive, Advanced Batteries
Houston, TX	Energy, Energy Services, Chemical Manufacturing, Shipbuilding, Oil Refining
Milwaukee, WI	Water Tech
New York, NY	IT, Biotech, Digital Media, Media, Healthcare, Advertising, Financial Products and Services
Northeast Ohio, OH	Additive Manufacturing, Glass, Solar Panels, Rubber and Polymers, LED, Flexible Electronics
Orlando, FL	Simulation, Optoelectronics
Phoenix, AZ	Air and Water Purification, Solar Technology
Pittsburg, PA	Robotics, Pollution Control Equipment, Medical Sciences
Portland, ME	Aircraft Products, Aircraft Parts
Portland, OR	Semiconductors, Electronics Manufacturing, CleanTech
Research Triangle Park, NC	Life Sciences, Advanced Technology
Rochester, NY	Materials/Imaging, Web Design, Digital X-rays, Military Optics Technology, Blood Analyzers
San Diego, CA	Biotech, Medical Sciences
Seattle, WA	Aircraft, Medical Diagnostics, Software
Silicon Valley, CA	Semiconductors, IT, Solar, Electric Vehicles
St. Louis, MO	Life Sciences
St. Paul/Minneapolis, MN	Medical Devices
Syracuse, NY	Clean Energy, Biosciences
Washington D.C.	IT, Government Contracting, Legal
Wichita, KS	Aircraft

industries stronger patent systems seem not to lead to more innovation. Patents are supposed to spread knowledge, by obliging holders to lay out their innovation for all to see, which they often fail to do because patent-lawyers are masters of obfuscation. This system has instead created a parasitic ecology of trolls and defensive patent-holders, who aim to block innovation, or at least to stand in its way unless they can grab a share of the spoils. Such parties obtain comprehensive patents for the purpose of stopping inventions, or appropriating the fruits of the inventions of others[197]. Patents should spur bursts of innovation; instead, they are used to lock-in incumbents' advantages.

Researchers argue that copyright law and especially patents prohibit the timely sharing of information, slow down research, discourage collaboration among scientists, and hold back new innovations. In 2011, Apple, Microsoft and other companies won Nortel Networks' six thousand patents worth $4.5 billion in auction; Google purchased Motorola for $12.5 billion, acquiring seventeen thousand patents; Microsoft purchased nine hundred and twenty-five patents from AOL for $1.1 billion; and Facebook bought six hundred and fifty patents from Microsoft for $550 million. The cost of the innovation that never takes place because of the flawed patent system is incalculable.

Abolition of patents though flies in the face of intuition that if you create a drug or invent a machine, you have a claim on your work. Inventors feel justifiably aggrieved if their ideas are stolen. Striking a proper balance between the claim of the individual and the interests of society is hard. The argument that the government should force the owners of intellectual property to share is especially strong – sharing ideas will not cause as much harm to the property owner as sharing physical property does. An imitator can reproduce an idea without depriving its owner of the original. Also, sharing brings huge benefits to society, partly from the wider use of the idea itself. Sharing also leads to extra innovation. Ideas are intangible, ideas overlap, and innovation is complex. Inventions depend on earlier creative advances. Innovation today is less about entirely novel breakthroughs, and more about the clever combination and extension of existing ideas.

A paper updated for the Federal Reserve Bank of St. Louis argues that patents are neither as good at rewarding innovation nor as helpful in propagating it as claimed[198]. Under-resourced patent-officers will always struggle against well-heeled patent-lawyers. Over the years, the regime is likely to fall victim to lobbying and special pleading. Hence a clear, efficient patent system is better than an elegant but complex one. In government as in invention, simplicity is strength. Forty to ninety percent of patents are never exploited or licensed out by their owners. Patents should come with a "use it or lose it" rule, so that they expire if the invention is not brought to market. They should also be easier to challenge without the expense of a full-blown court case.

---

[197]The Economist, August 08, 2015 - Intellectual Property: A Question of Utility

[198]Boldrin, Michele and Levine, David - The Case Against Patents, 2012

Patents should reward those who work hard on big, fresh ideas, rather than those who file the paperwork on a lark. A study in 2005 found that newcomers to the semiconductor business had to buy licenses from entrenched industry players for as much as $200 million. Patents also last too long. Protection for twenty years might make sense in the pharmaceutical industry, because to test a drug and bring it to market can take more than a decade.

Once subsidies and tax breaks are accounted for, American private industry pays for only about a third of the country's biomedical research. In return the patent system provides them with a great deal of income. Without the temporary monopoly that patents bestow, America might have saved three-quarters of its $210 billion bill for prescription drugs in 2005. A competitive patent-free market might have provided the same drugs for no more than $50 billion. The drug companies reckoned at that time they were spending $25 billion on R&D; the government was spending $30 billion on basic medical research. The money it would have been able to save buying drugs for Medicare and Medicaid in a patent-free world would have allowed the government to double that research spending, more than replacing industry's R&D, while still leaving $130 billion in public benefit. With America's prescription-drug bill now $374 billion, the opportunity looks all the greater, even though the companies now say they are investing $51 billion a year into R&D. Some economists have suggested encouraging teams of autonomous scientists to develop new breakthrough drugs by offering big prizes to those that succeed. Many drug startups see their exit strategy as being bought up for a billion dollars or so by a big pharmaceutical company when their projects start to look promising. Billion-dollar prizes would provide similar incentives.

When patents lag behind the pace of innovation, firms end up with monopolies on the building-blocks of new industry. The ability to patent has been extended from physical devices to software and stretches of DNA, to business processes and financial products. The fear of international competition has seen the system spread around the world, typically as the price that smaller or poorer nations pay for the access to developed countries' markets. One reason why talks on the proposed Trans-Pacific Partnership, a trade deal involving countries which produce forty percent of the world's economic output, ended inconclusively earlier in 2015 was the strong patent-protection Western countries wanted for biotech-based drugs. Patent filings tend to be carefully written so as to obscure how the patented idea works even from experts in the field. A counterpart to this defensive ploy is the filing of "submarine" patents: vague and speculative applications made by parties who then try, through various ploys, to keep the application from being granted until other people seem to be making progress on the technology in question. At that point the submarine surfaces with a view to demanding licensing fees[199]. If patents do not

---

[199] Johns, Adrian – Piracy: The Intellectual Property Wars from Gutenberg to Gates, 2010

hold many advantages, why do they persist and indeed multiply. In some industries and countries they have become a measure of progress in their own right – a proxy for innovation, rather than a spur. Another reason is self-defense. In much of the technology industry companies file large numbers of patents mostly to deter their rivals. Some studies have found "thickets" of patents that make it harder for all companies to launch new products. A top-to-bottom re-examination

---

Energy Efficiency Using Smart Windows
Thin-film Optical Retarders for Low-energy Smart Glass

This federally-funded research and development work from March 2012 to July 2015 in the amount of approximately $783,000 to a startup based in South Bend, Indiana was to develop low-cost smart-window technology. This R&D effort will utilize contemporary display industry fabrication and processing technologies to create unique large-area optical films. These films will be subsequently used to construct energy-efficient smart windows that modulate transmission or reflection of light on command. Windows, skylights, and other glazings made with this technology will have the ability to darken on command. Window-size prototypes will be designed, constructed and evaluated. Production, material costs, and prototype operation will be considered. Successful fabrication of these prototypes will enable smart windows to be manufactured in an electrochemically passive manner, simplifying their installation in existing windows, minimizing up-front costs, and ultimately reducing energy bills. The technology is also uniquely capable of being applied as an aftermarket or retrofit solution.

If this R&D effort is successful, it will result in potential savings of billions of dollars in energy costs in the United States alone, and a reduction of carbon footprint. Buildings are responsible for 70% of the electricity consumed in America. As part of a daylighting /natural heating strategy, smart window technologies have received much attention for their ability to reduce building energy consumption. Unfortunately, existing smart window products suffer from severe limitations in lifespan, scalability and cost. The technology that was developed here is a radically different approach to smart windows because instead of electrochemical processes, it utilizes stable films. This affords more chemical stability, longer life, better manufacturing scalability, power independence via manual operation, and lower costs to the consumer. This startup is in the business of the latest in window-blinds technology without the blinds. Two polarized films slide across each other to cut out the sun and provide perfect shade. This polarized film system is being designed into standard windows in homes. The amount of shade obtained is controlled from a smartphone or a controller mounted on the wall.

---

of whether patents and other forms of intellectual property protection actually do their job, and even whether they deserve to exist, is long overdue. Reductions in the duration of exclusive rights and differentiation between those rights for different sorts of innovation are possible, and could be introduced in steps for a number of years, allowing plenty of time for any ill effects to surface. Experiments with other forms of financing innovation could be run alongside the patent system. The call then is to undertake an operational overhaul of the United States Patent and Trademark Office along with its current dysfunctional incentive structures.

Future of Work, Future of Startups

In 2013, in the U.S., almost twenty-two million adults were unemployed, underemployed, or discouraged and no longer counted in the official statistics. Worldwide twenty-five percent of the adult workforce was unemployed, underemployed, or discouraged and no longer looking for work in 2011. The International Labor Organization reported that more than two hundred million were without work in 2013. In the U.S. between 1982 and 2002, steel production rose from 75 million to 120 million tons, while the number of steel workers declined from two hundred and eighty-nine thousand to seventy-four thousand. Between 1995 and 2002, twenty-two million manufacturing jobs were eliminated in the global economy while global production increased by more than thirty percent worldwide.

Although it is true that manufacturers that have long relied on cheap labor in their Chinese production facilities are bringing production back home with advanced robotics that are cheaper and more efficient than their Chinese workforces, what if technology permanently replaced a great deal of human work? What would society without jobs look like? While human capabilities remain the same, the capabilities of machines continue to expand exponentially – is any job therefore truly safe? Banishment of drudgery, expansive leisure, and almost limitless freedom may result. The disappearance of work would usher in a social transformation unlike any we have seen. By the nature of work, cultural breakdown would matter even more than the economic breakdown. Saving work is more important than saving any particular job. Industriousness has served as America's unofficial religion – the sanctity and preeminence of work lie at the heart of the country's politics, economics, and social interactions. What might happen if work goes away?

Unlike the agricultural industry, the industrial revolution, the services industry and globalization, when the total number of jobs always increased, what is looming is an era of technological unemployment when the total number of jobs declines steadily and permanently. Technology could exert a slow but continual downward pressure on the value and availability of work, where the expectation that work will be a central feature of adult life dissipates for a significant portion of society because of the ongoing triumph of capital over labor, the quiet demise of the working man, and the impressive dexterity of information technology. Since 2000, the number of manufacturing jobs has fallen by almost five million, or about thirty percent. The job market appears to be requiring more and more preparation for a lower and lower starting wage. It is forecasted that machines might be able to perform half of all U.S. jobs in the next two decades. Most jobs are still boring, repetitive, and easily learned; our newest industries tend to be the most labor-efficient: they just don't require many people. Sooner or later we will run out of jobs.

Founded by two technologists in a garage in Palo Alto in 1939, the giant created by Bill Hewlett and Dave Packard is remembered as Silicon Valley's original startup; in its heyday, its flair for innovation was unrivalled, and it devised technologies that shaped people's work and personal lives, from calculators to cameras to computers. Today it is regarded as a has-been: a reminder of Silicon Valley's past but not a beacon of its future. Steve Jobs chose to do a summer internship at HP. But today's engineers and aspiring entrepreneurs prefer to work at nimble startups where their ideas are more likely to be listened to, and the share price has more room to grow. Corporate life continues to involve dealing with largely anonymous owners, most of them represented by fund managers who buy and sell shares listed on a stock exchange. An enduring inefficiency of the market is aligning the interests of investors and owners.

The most distinctive aspect of America's vibrant startup sector is the way the ownership of companies is structured. In insurgent companies, the coupling between ownership and responsibility is tight, with founders, staff and backers exerting control directly; if this innovation spreads, it could transform the way companies work. The most interesting alternative to public companies is a new breed of high-potential startups that go by exotic names such as unicorns and gazelles. The central difference lies in ownership – rights and responsibilities are meticulously defined in contracts drawn up by lawyers to align interests and create a culture of hard work and camaraderie. New companies such as these also exploit new technology, which enables them to go global without being big themselves. Today startups can expand fast by buying services as and when they need them; they can incorporate online for a few hundred dollars, raise money from crowdsourcing sites such as Kickstarter, hire programmers from Upwork, rent computer-processing power from Amazon, find manufacturers on Alibaba, arrange payment systems at Square, and immediately set about conquering the world[200].

Vizio was the bestselling television brand in America in 2010 with just two hundred employees. WhatsApp persuaded Facebook to buy it for $19 billion despite having fewer than sixty employees and revenues of twenty million dollars. WeWork provides accommodation for startups, has eight thousand companies with thirty thousand workers in fifty-six locations in seventeen cities. Technology companies that list in America now do so after eleven years compared with four in 1999. Even when they do go public, technology entrepreneurs keep control through "A" class shares. The startup scene is dominated by a clique of venture capitalists with privileged access. That is true, yet ordinary people can invest in startups directly through platforms such as SeedInvest or indirectly through mainstream mutual funds such as T. Rowe Price, which buys into them during their infancy.

---

[200]The Economist, October 24th, 2015 - Reinventing the company: Entrepreneurs are redesigning the basic building block of capitalism

Once regarded as safe havens, the professions are now in the eye of the storm[201]. Machines are challenging the professions' two most important claims to being special: their ability to advance the frontiers of knowledge and their exclusive license to apply their expertise to an un-ordained laity. IBM and the Baylor College of Medicine have developed a system called KNIT - knowledge integration toolkit - that scans the medical literature and generates new hypotheses for research problems. Various bits of software regularly outperform legal experts in predicting the outcome of court decisions from patent disputes to America's Supreme Court. Nurses and "physician associates", equipped with computers and diagnostic tools, are doing more and more of the work once reserved for doctors. Every month one hundred and ninety million visit WebMD – more than visit regular doctors in America. Massive Open Online Courses are attracting millions of students. Judges and lawyers are increasingly resolving small claims through "e-adjudication". Machines have a bigger capacity for coping with complexity than humans. Expertise and empathy rarely come in the same package. New sub-disciplines will emerge, such as "knowledge engineers" who encode professional wisdom into software and various groups of para-professionals who then work out ways of applying this knowledge.

The demographics of employees are changing and so are employee expectations, values, attitudes, and styles of working. Conventional management models must be replaced with leadership approaches adapted to the future employee. Organizations must also rethink their traditional structure, how they empower employees, what they need to do to remain competitive in a rapidly changing world, how employees of the future will work, how managers will lead, and what organizations of the future will look like. Throughout the history of business, employees had to adapt to managers and managers had to adapt to organizations. In the future this is reversed with managers and organizations adapting to employees[202]. We will soon see automation high and low – robots performing surgery and behind fast-food counters, self-driving cars on streets, and drones dotting the sky replacing millions of drivers, warehouse stockers, and retail workers.

Some observers believe that the human capacity for compassion, deep understanding, and creativity are inimitable. As formal employment opportunities decline, the possible futures are one of consumption that is devoting one's freedom to simple leisure, communal creativity by seeking to build productive communities outside the workplace, and contingency meaning people fighting to reclaim their productivity by piecing together jobs in an informal

---

[201]Susskind, Richard and Susskind, Daniel - The Future of the Professions: How Technology Will Transform the Work of Human Experts, 2016

[202]Morgan, Jacob - The Future of Work: Attract New Talent, Build Better Leaders, and Create a Competitive Organization, 2014

economy. This could lead to the valorization of well-rounded resourcefulness, where entrepreneurship emerges out of necessity[203].

Big cities are highly sophisticated labor ecosystems. Technology trends could make it easier for people to start their own small-scale and even part-time businesses. One way to nurture fledgling ideas would be to build out a network of business incubators. We could take a lesson from Germany on job-sharing. Government could pay people to do something rather than nothing, such as caring for a rising population of elderly people. Local or federal government sponsorship of a national online marketplace of work for tasks that require empathy, humanity, or a personal touch; to imagine a flourishing post-work society – how will people discover their talents or the rewards that come from expertise, if they do not see much incentive to develop either? A future of less work still holds a glint of hope, because the necessity of salaried jobs now prevents so many from seeking immersive activities that they enjoy.

New Industry Creation

Major modern industries were created or hugely nurtured by taxpayer-paid research and development monies. These include the aerospace, biotech, nanotech, computer and internet, pharmaceutical, medical device, containerization, and agribusiness industries. In this regard, we can contemplate in the medium-term, 2015-2020, that America seriously consider creating a new public-private Department of Innovation, headed by the Secretary of Innovation who as Chair of the National Innovation Council, will like the national security advisor and the national economic advisor, report directly to the President, and will be the President's national innovation advisor. This new public-private partnering entity will be responsible and accountable for strategy, communication, execution, performance, and assessment of all federal small business research programs that fund and support startups leveraging knowledge-based innovation. It will also include the U.S. Patent and Trademarks Office.

The private-public partnership will facilitate the creation of a $10 billion New Industries Foundation consisting of multiple public-private venture funds specifically coordinated to identify and invest in technology platforms each capable of massive scaling and creation of new high-wage jobs in America. These technology sectors will be aligned with the already ongoing approximately $100 billion federally-funded research effort in the basic sciences that is likely to lead to whole new industries in America. Table 9.6 outlines a set of probable sunrise industries. These technologies are identified as strategic and critical to the U.S., and that current private institutional investors by

---

[203]Thompson, Derek - A World Without Work, The Atlantic, July/August 2015

themselves typically are loathe to invest in for reasons attributable to high-risk profiles and long incubation times required for exits.

Table 9.6   Probable Sunrise Industries

3D Printing	Human tissue, jet engine parts, toys, houses, ...
Advanced Manufacturing	Agile, adaptive, re-configurable production models; high-volumes to volumes of a few units
Advanced Materials	By design; biomimetics; intelligent materials; programmable matter; ultra-cheap housing
Advanced Medical Devices	Pacemakers; insulin pumps; operating room monitors; stents; defibrillators; surgical instruments; deep-brain simulators; drug delivery
Bio-computing	DNA for memory storage; bio-informatics; functional design
Bio-manufacturing and Industrial Microbiology	Non-fossil-based biodegradable industrial feedstock; pharmaceuticals; energy generation
Biotech	Inorganic biology; do-it-yourself biology; cheap vaccines; efficient drug discovery methodologies
Cities of the Future	Physical and cyber-infrastructure; electric transportation; autonomous vehicles
Clean Technology	Sustainability; low-carbon energy generation
Education	Global, cheap, accessible
Electric Transportation	Cars, public transport, airplanes, ships, rockets
Energy Conservation	Smart buildings; smart grids; off-grid energy generation; efficient, easily stored
Flexible Electronics Manufacturing	Organic electronics; ultra-thin, foldable displays; wearables; computing devices; printed books
Nanotechnology	Consumer products; programmable nanomachines; nanoscale measurement/instrumentation systems
Next-generation Internet	Wireless/mobile; remote access; big data; cloud; internet of things; IPv.6; social media
Next-generation Semiconductors	High-speed, parallel computing; stacked architectures; quantum computing
Nuclear Fusion	Distributed, personalized, small-size, non-polluting
Personalized Medicine	Genomics; proteomics; epigenetics; drug therapy
Precision Agriculture	Micro-irrigation; non-agricultural food production; agricultural biology; ultra-high-yield agriculture
Regenerative Medicine	Tissue engineering; organ printing
Robotics and Civilian Drones	Artificial intelligence; automation; autonomous technologies; precision agriculture; mapping; logistics
Solid-state Lighting	Rural electrification; integrated photonics; energy conservation
Space Tourism	Launch infrastructure; propulsion; reusability; hotels
Synthetic Biology	New drugs/vaccines; clothing; foods; artificial life
Water Technologies	Purification; recycling; desalination; conservation

The selection should typically also be based on global drivers such as climate change mitigation; global health, nutrition, and hygiene; alleviation of poverty; education; water and food; and governance and corruption. These funds will also be used to create the necessary infra-technologies such as the measurement techniques, sensor technologies, process controls, performance assurance, quality control, test-beds, pilot-scale production facilities, and standards for interfaces and protocols, along with the required technology strands meant to fill out platforms and/or supply-chains that would allow the manufacturing in America of products created by these new industries. We can also identify regional locations that provide incentive packages for manufacturing, and set up small business pre-competitive consortia and clusters in each technology area. The level, kind, and composition of financing will be technology-specific. In the timeframe 2030-2035, a perpetual $100 billion private-public New Industries Endowment should be crated to sustain ten evergreen funds, each a billion dollars a year technology-specific investment fund.

Technologies such as genomic medicine are at the forefront of a new customized approach to illness that treats each individual's affliction as an "orphan" disease. Genome engineering may help make porcine organs suitable for use in people. The clustered regularly interspaced short palindromic repeats/CRISPR-associated protein-9 (CRISPR/Cas9) system is a powerful method for disrupting your gene of interest, for targeted genome editing. It represents a significant improvement over these other next-generation genome editing tools, reaching a new level of targeting, efficiency, and ease of use. This system allows for site-specific genomic targeting in virtually any organism. Along with small interfering ribonucleic acid (siRNA) technology, data storage and cloud computing, peer-to-peer lending or social lending, games for education and other useful purposes, energy storage, micro-grid technology, geoengineering and bitcoin/block chain technology - the latest example of the unexpected fruits of cryptography, a shared, trusted public ledger that everyone can inspect, but which no single user controls - these are some of the technologies that can each create whole new industries in America.

Can we simply print money for innovation? For example, could the Federal Reserve, the U.S. government's banker, match Congress's overseas aid contributions, up to a certain level, by printing money? And do the same for innovation? Another possibility is the government issues perpetual bonds that pay no interest and the central bank then purchases these bonds in the required amount. Because the bonds never mature, they should not add to the national debt[204]. Other alternatives include creating money to buy bonds that are directly linked to technology development goals. The International Finance Facility for Immunization, for example, uses long-term donor pledges to issue highly-rated bonds to raise funds for vaccinations. Creative financial instruments to serve as new technology-specific investment vehicles such as public-private partnerships, crowd-funding, social

---

[204]Jackson, Andrew and Dyson, Ben – Modernizing Money: Why our Monetary System is Broken and How it Can be Fixed, 2013

impact bonds, innovation bonds, job creation bonds, vouchers, warrants, prizes, foreign direct investment, and philanthropy should be explored.

Healthcare Information Technology
Efficient Comparative Effective Research Tools in Real Time

This publicly-funded R&D work from August 2012 to July 2017 in the amount of approximately $1.1 million will help this startup based in Cupertino, California to enable comparative effectiveness research (CER), evidence-based precise and personalized medicine by researchers utilizing both clinical and complex biomolecular data. A feasibility study was completed working with customers to determine the extent to which the tools the startup was creating could be utilized in practice. Implementation of these tools in real-world healthcare environments was undertaken to create systemic improvements in healthcare by improving tools available for self-evaluation by hospital administrators and clinicians alike. The R&D effort involves the development of care-improvement algorithms for pediatric intensive care units for patients with sepsis and congenital heart disease; a reporting engine that mines electronic medical records to create CER reports for patients; an easy-to-use enterprise-level reporting engine to check compliance with national standard healthcare quality metrics and thus create new metrics for care improvement; and to create a personalized medicine reporting system based on next-generation sequencing data.

This R&D effort, if successful, will capitalize on two trends in healthcare - digitizing patient clinical data and the increasing use of molecular sequencing and microarray data. This technology will improve patient outcomes. Leveraging historical patient data in CER to eliminate ineffective therapies from the healthcare system, coupled with utilizing genetic information to create a more personalized model of healthcare, will focus healthcare dollars on effective therapies optimized for the individual. This tool would be useful to clinicians, researchers, pharmaceutical companies, and insurance companies due to increased quality and cost savings. This product suite is a fully integrated, open platform that completes the entire translational medicine/ continuous learning workflow from data aggregation, to retrospective analysis to prospective clinical application. By making significant investments to collect and store vast amounts of healthcare data, healthcare providers have taken the first crucial step on the path to becoming a continuous learning organization. These disconnected data repositories are an untapped gold mine continually replenished with evidential lessons that can improve clinical quality and patient outcomes while fueling the organization's continuous learning loop. The true return-on-investment of this data collection can be realized when it is organized to continuously and systematically discover improvements and translate them into clinical practice. Additionally, this platform effectively and efficiently translates discoveries arising from basic science, clinical research, clinical trials, and population studies into everyday clinical practice through informed decision support.

The B-Corp concept says that business, the most powerful man-made force on the planet, must create value for society, not just shareholders. Systemic challenges require systemic solutions and the B-Corp movement offers a concrete, market-based, and scalable solution. In the U.S., the "benefit corporation" is a new business model that is attempting a makeover of the conventional capitalist corporation to allow it to be more agile and able to maneuver in the hybrid world of markets and the collaborative commons that could displace industrial capitalism when the peer-to-peer economic and social practices of the internet are extended to energy, logistics, and material fabrication. Other ideas include the economic growth bank, the national infrastructure fund, a national innovation fund, and the use of the Small Business Investment Corporation and the Export-Import Bank for small business pilot-scale and full-scale manufacturing in America. There are also advocates for a new kind of bank – the development bank model[205].

Our country is laden with leaden fortunes, basking in languorous investments, some of whose possessors can be brought forth to advance our society in a way that would invigorate these generous donors with fresh significance[206]. There are trillions of dollars held by upper-income Americans in what might be called "inert" investments, many in market funds, savings banks, and treasuries, bringing in a fraction of 1% in interest. Revenue-based financing, a new type of startup financing wherein an investment firm invests a hundred thousand to a million dollars in a startup and the startup pays it back with monthly earnings rather than through an initial public offering or another exit type. A venture capital firm invests in a company or team, while a revenue-based firm invests in a product; a revenue-based firm would not take equity in the start-up. When it is based on revenue, it builds more flexibility into the payment system. A possible negative for investors is that returns would be relatively modest compared with those a venture capital firm might see with a high-flying portfolio company, that is, if it finds one! Also, as they pay investors back, entrepreneurs have less capital to reinvest in their company.

Impact investing, exploring the intersection of lucre and idealism, invests money into companies that seek to advance human well-being and still make a profit. Many family offices, foundations, endowments, and pensions have already committed money into impact investments or are considering it. Peer-to-peer lending or social lending is a growing phenomenon - by the end of 2013, Kickstarter had fostered fifty-one thousand projects with a forty-four percent success rate; the projects had raised more than $871 million in 13 categories – art, dance, design, fashion, films and video, food, games, music, photography, publishing, technology, and theater. There are other crowdfunding platforms like Indiegogo, Early Shares, Crowdfunder, Fundable, and Crowdcube. Passage of the Jumpstart Our Business Startups (JOBS) Act in 2012 allows small businesses to raise as much as one million dollars in investments annually from the general public via crowdfunding platforms.

---

[205]Ross, Carne - The Leaderless Revolution: How Ordinary People Will Take Power and Change Politics in the 21st Century, 2012

[206]Nader, Ralph – Unstoppable: The Emerging Left-Right Alliance to Dismantle the Corporate State

Economic paradigms are just human constructs, not natural phenomena. The near-zero marginal cost phenomena has already wreaked havoc on the publishing, communications, and entertainment industries as more and more information is being made available nearly free to billions of people. It is beginning to affect other commercial sectors, including renewable energy, 3D printing in manufacturing, and online higher education. How would economic life proceed in a world where most goods and services are nearly free, profit is defunct, property is meaningless, and the market is

---

Control with Unknown Muscles
Harnessing Vestigial Neuromuscular Biosignals

This taxpayer-funded research and development work from September 2013 to February 2017 in the amount of approximately $900,000 to a startup based in Los Angeles, California is to explore the extent to which signals from the vestigial muscles around the ears can be repurposed as a new human ability. Higher primates have not needed these muscles for millions of years, they are no longer functionally significant in humans, yet they exist in nearly everyone and are controlled by nerves that come directly from the brainstem, thereby bypassing the spinal cord to remain functional after even the most severe spinal cord injuries. The ear muscles, or peri-auricular muscles, are part of a vestigial human system that evolved to orient the ears towards sounds. It thus has the potential to serve as an innate and intuitive way to direct a computer cursor, operate electronics, or drive a motorized wheelchair. This R&D effort involves the development of an optimized headset controller device, along with the testing and evaluation of this new control modality in a variety of real-world applications that would be useful for both paralyzed and able-bodied people.

The impact of this work is expected to be significant in the human assistive technology space. Limitations of mobility and interpersonal interaction are primary factors determining functional independence and quality of life in people with physical disabilities, yet existing assistive devices remain cumbersome and usurp the individual's few remaining still-functioning motor systems. This innovation could bring a new controller to those with paralysis or limb loss, and has the potential to change how such persons interface with their environment. Users would, in essence, develop a new mode of command output that can be operated intuitively, invisibly, and wirelessly. The startup's team of clinicians, researchers, computer scientists, and engineers has developed technology to harness a long lost neuromuscular system. The startup's headset is a next-generation wearable controller, utilizing proprietary sensor and software technology. The system measures the electromyography signals generated by the muscles around the ears and wirelessly transmits those signals to target devices as commands.

---

superfluous[207]? Patents and copyrights, for example, thrive in an economy of scarcity but are useless in an economy organized around abundance. Whether at a certain stage of technological development the very success of the system would become a shackle to its further advance. Keynes had it correct when he speculated about

---

[207]Rifkin, Jeremy – The Zero Marginal Cost Society: The Internet of Things, the Collaborative Commons, and the Eclipse of Capitalism, 2015

technological unemployment, that is unemployment due to our discovery of means of economizing the use of labor outrunning the pace at which we can find new uses for labor. A future in which

---

Learn from Nature to Develop New Organisms
Proteolysis-based Tools for Metabolic Engineering

This federally-funded research and development work from April 2013 to September 2017 in the amount of approximately $815,000 to a startup based in Boston, Massachusetts is to engineer microbes for the cost-effective production of specialty chemicals. Currently, engineered microbial strains bear mutations that increase the production of chemicals of interest by inhibiting the cell's ability to produce off-pathway chemicals. These "loss-of-function" mutations are critical as they effectively channel the cell's metabolic flux toward the product of interest. This boosts the production efficiency and eases downstream purification by eliminating the accumulation of undesirable but chemically-similar contaminants. Unfortunately, these mutations may also decrease the fitness of the cells and, as a result, the growth media must be supplemented with costly nutrients. This R&D effort will assess the feasibility of applying novel regulated proteolysis technology - the breakdown of proteins or peptides into amino acids by the action of enzymes - to simultaneously direct maximal metabolic flux toward the target chemical of interest while avoiding the need to supplement the growth media. The success of this R&D will result in the provisioning of a stable and cost-effective fermentative production route to a specialty chemical. Fermentative production of chemicals offers many advantages over traditional petrochemical or extraction-based production processes. Petrochemical production maintains the nation's reliance on an unsustainable fossil fuel-based feedstock. Chemical production via extraction from plant materials also has ecological challenges. The process often uses toxic solvents, and may rely on unsustainable farming practices for many plants that are not traditional food crops. Engineered microbes fermented on sugar feedstock produced using high-efficiency agricultural practices offer a stable alternative for producing specialty chemicals, both in terms of supply and price.

The startup was founded with the mission to make biology easier to engineer. The company has built the world's first organism engineering foundry thus expanding the role of biology in our world. Nature's biodiversity holds answers to challenges in health, energy, food, and materials. The company is also working with professional perfumers to create the world's first designer cultured rose extract. For the first time perfumers will work directly with organism designers to design a new fragrance. Rose breeders have altered wild rose species to be more fragrant, but until now have not been able to control the exact makeup of the rose essence. Humans have been culturing foods for millennia. Beer and wine brewing using yeast cultures began about 8,000 years ago and cultured foods then expanded to cheese, yogurt, soy sauce, sauerkraut, and breads. The company has so far licensed eight of its organisms for the production of cultured ingredients. The integration of advanced software, robots, and biology allow this startup to rapidly design, build, and test organisms until they meet customer specifications.

---

machines would produce an abundance of nearly free goods and services, liberating the human race from toil and hardships and freeing the human mind from a preoccupation with strictly pecuniary interests to focus more on the "arts for life" and the quest for transcendence.

We need a new theoretical economics grounded in the laws of thermodynamics. Standard capitalistic theory is virtually silent on the dissoluble relationship between economic activity and the ecological constraints imposed by the laws of energy. In classical and neoclassical economic theory, the dynamics that govern Earth's biosphere are mere externalities to economic activity – small, adjustable factors of little real consequence to the working of the capitalistic system as a whole. It is urgently incumbent upon humanity to adopt a circular economy, one in which everything is recycled and reused and nothing is sent to the landfill before its time. Another example is provided by micro-currencies - there are more than four thousand kinds in circulation around the world; many of them are based on the labor-time one person gives to another in making a good, repairing an item, or performing a service; hours are stored in a time-bank, just like cash, and exchanged for other hours of goods and services. The time-bank is inspired by people giving blood at a blood bank, based on a core principle that underlies the social economy – reciprocity; some community currencies are also employed, in part, to prevent wealth from leaking out of the community.

As always, it is not necessary for government to shoulder the entire burden, but only to facilitate and underwrite the process. If democracy and economic freedom in America is to endure, we must find ways to resist the allure of technology by cultivating a deeper appreciation of the natural world. Technology has greatly improved our quality of life but it has also raised concerns and issues that give us pause. It so pervades our existence that it almost feels like a natural part of life, as though we have been using it forever, even accepting its imperfections as we would of nature itself. We often leverage technology to increase knowledge and shape our thinking but the manner by which we use technology may influence human character. We acquire a piece of technology for leisure, convenience, knowledge, and perhaps to shape our thinking but the very act of acquisition may by itself influence our character for the worse. Technology allows us to sometimes behave in unethical ways, at the same time its use may also have unethical consequences. In general, Americans believe knowledge is power and science is valuable if it can be usefully applied.

Acknowledgement

It is a privilege and an honor to be given this opportunity to work as a public servant in the United States federal government. This book is a culmination of over a dozen years of experience with the small business innovation research program at the National Science Foundation. It would not have been possible to complete it without the active support and assistance of my colleagues who offered valuable insights, advice, and suggestions.

I would like to thank and acknowledge the presence in my life of my inspirational muse, my infinitely supportive and patient wife.

List of Tables

Table 2.1	Employment Numbers - Select U.S. Firms	47
Table 3.1	Public Sector-seeded Industries	57
Table 3.2	Sources of U.S. R&D Funding	58
Table 3.3	NIH Institutes and Centers	61
Table 3.4	DOE National Labs	64
Table 3.5	NASA Centers and Facilities	67
Table 3.6	NSF-funded Engineering Research Centers	70
Table 4.1	Listing of NSF-seeded Technology	77
Table 4.2	Invented in U.S. Now Mostly Made Abroad	78
Table 4.3	SBIR by the Numbers	83
Table 4.4	NSF SBIR-seeded Fantastic Four	84
Table 5.1	U.S. Venture Capital Industry	110
Table 5.2	General Investing Considerations	111
Table 5.3	Early-stage Investments by VC Firms	112
Table 5.4	Top Investors in High-tech	113
Table 5.5	Startup Support Networks	122
Table 6.1	Typical Startup Needs	128
Table 6.2	Network Build-out Mechanisms	131
Table 6.3	Partial List of Potential Partners	137
Table 6.4	Y Combinator Greatest Hits	142
Table 6.5	Biggest Pharmaceutical M&A	143
Table 6.6	Common Venture Capital Investment Criteria	145
Table 6.7	Typical Startup Business Models	147
Table 7.1	Startup Growth Scenarios	155
Table 7.2	Biotechnology Value-chain	160
Table 7.3	Possible Set of Platform Technologies	161
Table 8.1	Potential Advanced Manufacturing Sectors	182
Table 9.1	Global Grand Challenges	188
Table 9.2	Greatest 20th Century Engineering Achievements	189
Table 9.3	Federally-funded Startups Portfolio Evolution	190
Table 9.4	Top U.S. Exports	199
Table 9.5	Technology Cluster Examples	201
Table 9.6	Probable Sunrise Industries	209

Photo/Illustration Credits

front inside jacket - gettyimages.com

Startup Profiles Listing

Introduction

Location	Sector	Technology	
Eden Prairie, MN	Regenerative Medicine	No Waiting for an Organ Transplant! Cardiac-derived Patch for Heart Disease	4
Pittsford, NY	Education Tech	Martha Madison's Marvelous Machine: New Learning Pathways	5
Alameda, CA	Battery Tech	Rethink Energy Storage: Printable Non-lithium Battery Technology	9
Charlotte, NC	Water Purification	Water is the New Oil: Deep-ultraviolet Technology for Water Disinfection	10
Salt Lake City, UT	Smart Lighting	Quantum Dots for Lighting: Low-cost Manufacturing of Semiconductor Nanocrystals	13

Chapter 1   Society and Technology

Location	Sector	Technology	
Austin, TX	Life Sciences	Public Health, Entrenched Interests: Flexible Plastic Packaging without Estrogenic Activity	23
Mountain View, CA	Precision Agriculture	Eliminate Back-breaking Work: Machine Learning Applied to Robust Crop and Weed Detection	24
Cambridge, MA	Food	Technology Looking for Problem: Thermal Energy Storage to Prevent Food Spoilage	27
Sunnyvale, CA	Education Tech	Make Education More Accessible: Simulated Patients for Medical Training	28
Mountain View, CA	Medical Devices	Reduce Kidney Disease Infections: Novel Peritoneal Dialysis Catheter	30
Palo Alto, CA	Biodegradable Materials	Reduce Environmental Degradation: Waste Methane Gas to Biodegradable Biopolymer	33

Chapter 2   Innovation Economy

Location	Sector	Technology	
Minneapolis, MN	Instrumentation	Material Science Challenges: Nanomechanical Characterization	48
Greenville, IN	Aerospace	Shooting for the Stars: Multipurpose Variable-gravity Platform	49

San Diego, CA	Recycling Tech	Recycle Used Cellphones: Kiosk Hardware for Automated Inspection	50
Kirkland, WA	Aerospace	Rockets into Space: Next-generation Missile and Launch Systems	51
Miami, FL	Medical Devices	Regenerative Biologic Therapies: Ocular Hygiene Solutions	52

Chapter 4   **Innovation Research**

Covington, KY	Pharmaceuticals	Innovative Cures for Cancer: Novel Mechanism for Targeting and Eliminating Tumor Cells	79
Lincoln, NE	Environmental Sciences	Portable Photosynthesis System: Simultaneous Analysis of Carbon Dioxide and Water Vapor	80
Seaford, DE	Energy Efficiency	Hybrid Water Heater: Electrochemical Compressor for 50-Gallon Residential Hot Water System	81
San Jose, CA	Renewable Energy	Reduce Cost of Solar Energy: Automated Utility-scale Photovoltaic Panel Installation System	82
Halethorpe, MD	Data Analytics	Maps, Data, and People: Big Data Analysis on Interactive Maps	85
Madison, WI	Space Exploration	Low-Temperature Device for Space: High-frequency Single-stage Pulse Tube Cryocooler	86
Hattiesburg, MS	Materials Tech	Improve Traditional Materials: Technology Based on Nanostructured Chemicals	87
San Francisco, CA	Wireless Tech	Cryogenic Radio Frequency Tech: Device Packaging and Control Electronics	88
San Francisco, CA	Synthetic Biology	Spider Silk without Spiders: Production of Synthetic Spider Silk Fibers	89
Corvallis, OR	Wearables	No Batteries Required: Wearable Energy Harvester and Wireless Sensor Networks	90

Chapter 5   **Infrastructure**

Boston, MA	Wireless Tech	Affordable High-speed Internet: Low-cost Transparent Wireless Mesh Network Node	93

Cupertino, CA	Semiconductors	One-transistor Memory Device: Chips for High-performance, Low-power Applications	96
Troy, NY	Biomaterials	Natural, Renewable, Biodegradable: Materials from Agricultural Byproducts and Mushrooms	99
St. Louis, MO	Chip Design	Electronic Design Automation: Simulate Synchronizer Behavior in System-on-Chip Design	101
Santa Barbara, CA	Instruments	Revealing Nanoscale-level Properties: Infrared Chemical Spectroscopy	104
Detroit, MI	Manufacturing	More Efficient Manufacturing: Minimum Quantity Lubrication for Metal Forming Applications	108
Waltham, MA	Software	Emotion-reading Machines: Cloud-enabled Analysis of Facial Affect	114
Evanston, IL	Display Tech	Rediscover the Sense of Touch: Force Modulation for Haptic Touchscreens	117
Champaign, IL	Aerospace	Flight Control Software: Scalable Semantics-Based Verification	120
Philadelphia, PA	Robotics	Reduce Constraints on Daily Living: Active Wheelchair Driving Aid for Independent Living	121

Chapter 6   **Engender Demand**

Menlo Park, CA	Software	Next-generation Internet Search: Automated Search to Repurpose Approved Drugs	129
Bellefonte, PA	Medical Devices	From Concept to Regulatory Approval: Instrument for Minimally Invasive Procedures inside Active Organs	134
Lebanon, NH	Biotech	Cancer Therapeutics: Building Discoveries into Investment-grade Startups	146

Chapter 7   **Consequence**

Waltham, MA	Software	Cybersecurity Performance: Information Security Risk Rating	151
Ithaca, NY	Organic Electronics	Flexible Displays: Manufacturing Organic Electronic Devices	153

Richmond, CA	Robotics	I Can Walk Again! Did You Say IPO? Powered Knee-ankle Prosthetic System	156
Port Hueneme, CA	Biotech	Treasures from the Ocean: Valuable Medical Resource Bottleneck	157
Forster City, CA	Instruments	Soot Sensor to Micro-spectrometer: Innovative Use of Electron Spin	159
West Lebanon, NH	Energy Storage	For a Cleaner, More Efficient Grid: Low-cost Energy Storage	162

Chapter 8 **Make in America**

Atlanta, GA	Display Tech	Next Dimension in Touch: Force Touch Stylus Experience	174
Cambridge, MA	Chip Design	Mobile Electronic Devices: Smaller, Thinner Chips for Power Management	175
Westminster, CO	Renewable Energy	Boost Tower Production Throughput: Tapered Spiral Welding for Wind Turbine Towers	178
Atlanta, GA	Medical Devices	Miracle Materials: Shape-memory Device to Heal Muscle Atrophy	179
Dallas, TX	3D Printing	Additive Manufacturing: Facilitate 3D Printing of Industrial Parts	183
Philadelphia, PA	Manufacturing	Graphene, Emerging Super Material: Roll-to-roll Production of Uniform Graphene Films	184

Chapter 9 **New Industry Creation**

Rayleigh, NC	Energy	Modernize America's Electricity Grid: Solid-state Transformer as Energy Router	191
Pittsburgh, PA	Biomimetics	Biomimetics for Gripping Materials: Manufacturing of Bio-inspired Polymer Fiber Arrays	193
South Bend, IN	Smart Buildings	Energy Efficient Smart Windows: Thin-film Optical Retarders for Low-energy Smart Glass	204
Cupertino, CA	Healthcare IT	Healthcare Information Technology: Efficient Comparative Effective Research Tools in Real Time	211

| Los Angeles, CA | Brain-computer Interfaces | Control with Unknown Muscles: Harnessing Vestigial Neuromuscular Biosignals | 212 |
| Boston, MA | Synthetic Biology | Learn from Nature: New Proteolysis-based Tools for Metabolic Engineering | 214 |

References

Introduction

1. Hayek, Freidrich - The Road to Serfdom, 1944
2. Stiglitz, Joseph - Globalization and Its Discontents, 2003
3. Habermas, Jurgen - The Philosophical Discourse of Modernity, 1985
4. Speth, James - The Bridge at the Edge of the World: Capitalism, the Environment, and Crossing from Crisis to Sustainability, 2008
5. Kuhn, Thomas - The Structure of Scientific Revolutions, 3rd Edition, 1996
6. McLuhan, Marshall - The Mechanical Bride: Folklore of Industrial Man, 1951
7. Note: In each chapter, snapshots of publicly-funded startups are featured to showcase the variety of technologies supported. The underlying theme for the five startups presented in this chapter is how such fledgling companies can and do provide solutions to global challenges such as health, education, energy, water, and lighting respectively.
8. Rischard, Jean-Francois - High Noon: Twenty Global Problems, Twenty Years to Solve Them, 2002
9. Cole, Jonathan - The Great American University: Its Rise to Preeminence, Its Indispensable National Role, Why It Must Be Protected, 2010
10. Drucker, Peter - Innovation and Entrepreneurship, 2006

Chapter 1

11. Wilson, Edward - The Future of Life, 2003
12. Stokes, Donald - Pasteur's Quadrant: Basic Science and Technological Innovation, 1997
13. Holton, Gerald - Thematic Origins of Scientific Thought: Kepler to Einstein, 1988
14. Heidegger, Martin - Being and Time, 2008
15. Kuhn, Thomas - The Structure of Scientific Revolutions, 1996
16. Arthur, Brian - Nature of Technology: What It Is and How it Evolves, 2009
17. Kelly, Kevin - What Technology Wants, 2010
18. Blake, William - The Marriage of Heaven and Hell, 1793
19. Marx, Karl - The Communist Manifesto, 1948
20. Bacon, Francis - Of the Proficience and Advancement of Learning, Divine and Human, 1605
21. McClellan, James and Dorn, Harold - Science and Technology in World History: An Introduction, 2006
22. Hughes, Thomas - Human-Built World: How to Think about Technology and Culture, 2005
23. Ellul, Jacques - The Technological Society, 1967
24. Winner, Langdon - The Whale and the Reactor: A Search for Limits in an Age of High Technology, 1989
25. Mumford, Lewis - The Myth of the Machine, 1967
26. Schlesinger, Arthur - The Disuniting of America: Reflections on a Multicultural Society, 1998
27. Hjorth, Linda et al - Technology and Society: Issues for the 21st Century and Beyond, 2007
28. Wilson, Edward - Consilience: The Unity of Knowledge, 1999
29. Lanier, Jaron - Who Owns the Future? 2013
30. Note: The underlying theme for the five instances presented in this chapter is how technologies created by startups funded by the U.S. federal government can provide workable solutions to societal problems such as public health; agricultural worker shortage; food spoilage and child malnutrition; environmental degradation; and universal education.
31. Kidder, Tracy - The Soul of a New Machine, 1981
32. Reich, Robert - Aftershock: The Next Economy and America's Future, 2010
33. Florida, Richard - Flight of the Creative Class: The New Global Competition for Talent, 2007

34. Gertner, Jon - The Idea Factory: Bell Labs and the Great Age of American Innovation, 2013
35. Watson, James - The Double Helix: A Personal Account of the Discovery of the Structure of DNA, 2001
36. Venter, Craig - Life at the Speed of Light: From the Double Helix to the Dawn of Digital Life, 2013
37. Turing, Alan - Computing Machinery and Intelligence, 1950
38. Chardin, Pierre Teilhard de – The Phenomenon of Man, 2008
39. Kurzweil, Ray - The Age of Spiritual Machines: When Computers Exceed Human Intelligence, 1999
40. Nielson, Donald - A Heritage of Innovation: SRI's First Half Century, 2006
41. Moravec, Hans - Robot: Mere Machine to Transcendent Mind, 1999
42. Safranski, Rüdiger and Osers, Ewald - Martin Heidegger: Between Good and Evil,1999
43. Schumpeter, Joseph - Essays: On Entrepreneurs, Innovations, Business Cycles, and the Evolution of Capitalism, 1989
44. Ruskin, John - The Stones of Venice: The Sea-Stories, 1853

Chapter 2

45. Phelps, Edmund - Mass Flourishing: How Grassroots Innovation Created Jobs, Challenge, and Change, 2013
46. Autor, David and Katz, Lawrence - Grand Challenges in the Study of Employment and Technological Change, 2010
47. Lester, Richard and Hart, David - Unlocking Energy Innovation: How America Can Build a Low-Cost, Low-Carbon Energy System, 2011
48. Baumol, William et al - Good Capitalism, Bad Capitalism, and the Economics of Growth and Prosperity, 2009
49. Sachs, Jeffrey - Common Wealth: Economics for a Crowded Planet, 2009
50. Elkus, Richard - Winner Take All: How Competitiveness Shapes the Fate of Nations, 2008
51. Prestowitz, Clyde - Three Billion New Capitalists: The Great Shift of Wealth and Power to the East, 2005
52. Atkinson, Robert and Ezell, Stephen - Innovation Economics: The Race for Global Advantage, 2012
53. Landes, David - The Wealth and Poverty of Nations: Why are Some So Rich and Some So Poor, 1999
54. Smith, Adam - An Inquiry into the Nature and Causes of the Wealth of Nations, 1759
55. Keynes, John - The General Theory of Employment, Interest, and Money, 1965
56. Schumpeter, Joseph - Capitalism, Socialism and Democracy, 1942
57. Arthur, Brian - The Nature of Technology: What It Is and How It Evolves, 2009
58. Bhide, Amar - The Venturesome Economy: How Innovation Sustains Prosperity in a More Connected World, 2008
59. Solow, Robert - A Contribution to the Theory of Economic Growth, 1956
60. Romer, Paul - Endogenous Technological Change, Journal of Political Economy, 1990
61. Coleman, James - Social Capital in the Creation of Human Capital, 1988
62. Aldrich, Howard and Martinez, Martha - "Entrepreneurship as Social Construction: An Evolutionary Approach." in Zoltan J. Acs and David B. Audretsch (Editors) - Handbook of Entrepreneurship Research: An Interdisciplinary Survey and Introduction, 2003
63. Thornton, Patricia and Flynn, Katherine - Entrepreneurship, Networks, and Geographies, 2003
64. Bathelt, Harald and Glucker, Johannes - The Relational Economy: Geographies of Knowing and Learning, 2011
65. Katz, Bruce and Bradley, Jennifer - The Metropolitan Revolution: How Cities and Metros are Fixing Our Broken Politics and Fragile Economy, 2013
66. Gore, Al - The Future: Six Drivers of Global Change, 2013

67. Porter, Michael - The Competitive Advantage of Nations, 1998
68. Florida, Richard - The Flight of the Creative Class: The New Global Competition for Talent, 2007
69. Flichy, Patrice - Understanding Technological Innovation: A Socio-Technical Approach, 2007
70. Wulf, William - The Innovation Ecology, 2008
71. Sagan, Carl - Pale Blue Dot: A Vision of the Human Future in Space, 1997
72. Bush, Vannevar - Science: The Endless Frontier, 1945
73. Stokes, Donald - Pasteur's Quadrant: Basic Science and Technological Innovation, 1997
74. Griliches, Zvi - R&D and Productivity: The Econometric Evidence, 1998
75. Chesbrough, Henry - Open Innovation: The New Imperative of Creating and Profiting from Technology, 2005
76. National Academy of Sciences - Is America Falling off the Flat Earth?, 2007
77. National Academy of Sciences - Rising Above the Gathering Storm: Energizing and Employing America for a Brighter Economic Future, 2007
78. Fitzpatrick, Erika - Innovation America: A Final Report (National Governor's Association), 2007
79. Friedman, Thomas - The World is Flat: A Brief History of the Twenty-first Century, 2005
80. The Economist: A Special Report on Innovation in Emerging Markets (Masters of the New Management), 2010
81. Christensen, Clayton - The Innovator's Dilemma: When New Technologies Cause Great Firms to Fail,1997
82. Hippel, Eric - Sources of Innovation, 1994
83. Hippel, Eric - Democratizing Innovation, 2006
84. Schramm, Carl - The Entrepreneurial Imperative: How America's Economic Miracle Will Reshape the World (and Change Your Life), 2006
85. Keeble, David and Wilkinson, Frank - Collective Learning and Knowledge Development in the Evolution of Regional Clusters of High Technology SMEs in Europe, 1999
86. Knight, Frank and McClure, John - Risk, Uncertainty, and Profit , 2009
87. Drucker, Peter - Innovation and Entrepreneurship, 2006
88. Note: The five startups profiled in this chapter feature examples of job creation by publicly-funded high-tech small businesses.
89. Kao, John - Innovation Nation: How America is Losing Its Innovation Edge, Why It Matters, and What We Can Do to Get It Back, 2007
90. Clifton, Jim - The Coming Jobs War, 2011
91. Clinton, Bill - Back to Work: Why We Need Smart Government for a Strong Economy, 2011
92. Fletcher, Ian - Free Trade Doesn't Work: What Should Replace It and Why, 2011
93. Kauffman Foundation - Roadmap for an Entrepreneurial Economy, 2006
94. Friedman, Thomas - Hot, Flat, and Crowded: Why We Need a Green Revolution - and How It Can Renew America, 2009

Chapter 3

95. Smith, Adam - An Inquiry into the Nature and Causes of the Wealth of Nations, 1776
96. Kennedy, Joseph - The Sources and Uses of U.S. Science Funding, 2012
97. Polanyi, Karl - The Great Transformation: The Political and Economic Origins of Our Time, 1944
98. Mazzucato, Mariana - The Entrepreneurial State: Debunking Private v. Public Sector Myths, 2013
99. Keynes, John - The End of Laissez-Faire, 1926
100. www.battelle.org - 2014 Global R&D Funding Forecast, 2013
101. Auerswald, Philip and Branscomb, Lewis - Valleys of Death and Darwinian Seas, 2003
102. www.nih.gov
103. www.energy.gov

104. www.nasa.gov
105. www.nsf.gov
106. National Science Board Science and Engineering Indicators, 2015
107. www.darpa.mil

Chapter 4

108. Goldfarb, Zachary - Relax. The Economy Will be Just Fine, The Washington Post, February 16, 2014
109. Brynjolfsson, Erik and McAfee, Andrew - The Second Machine Age: Work, Progress, and Prosperity in a Time of Brilliant Technologies, 2014
110. The Economist - Briefing: The Future of Jobs; Special Report: Tech Startups, January 18, 2014
111. Acs, Zoltan and Audretsch, David - Innovation and Small Firms, 1990
112. Block, Fred and Keller, Matthew - Where Do Innovations Come From?, 2008
113. Senor, Dan and Singer, Saul - Start-up Nation: The Story of Israel's Economic Miracle, 2011
114. Bhide, Amar - The Origin and Evolution of New Businesses, 2000
115. Reis, Eric - The Lean Startup: How Today's Entrepreneurs Use Continuous Innovation to Create Radically Successful Businesses, 2011
116. Eskesen, Anne - Innovation Development Institute, 2013

Chapter 5

117. Note: For each chapter subsection describing a spoke, a startup example illustrates the particular spoke.
118. Schumpeter, Joseph - Capitalism, Socialism and Democracy, 1942
119. Schramm, Carl - The Entrepreneurial Imperative: How America's Economic Miracle Will Reshape the World (and Change Your Life), 2006
120. Baumol, William - The Micro-theory of Innovative Entrepreneurship (The Kauffman Foundation Series on Innovation and Entrepreneurship), 2010
121. Drucker, Peter - Innovation and Entrepreneurship, 2006
122. Kuznets, Simon - Inventive Activity: Problems of Definition and Measurement, 1962
123. Jaffe, Adam et al - Patents, Citations, and Innovations: A Window on the Knowledge Economy, 2005
124. Bessen, James and Meurer, Michael - Patent Failure: How Judges, Bureaucrats, and Lawyers put Innovators at Risk, 2009
125. Heller, Michael - The Gridlock Economy: How Too Much Ownership Wrecks Markets, Stops Innovation, and Costs Lives, 2008
126. Choate, Pat - Hot Property: The Stealing of Ideas in an Age of Globalization, 2007
127. Jaffe, Adam and Lerner, Josh - Innovation and Its Discontents: How Our Broken Patent System is Endangering Innovation and Progress, and What to Do About It, 2004
128. Rivette, Kevin and Kline, David - Rembrandts in the Attic: Unlocking the Hidden Value of Patents, 1999
129. Phelps, Edmund - Economic Justice and the Spirit of Innovation, 2009
130. Hippel, Eric - Democratizing Innovation, 2006
131. Bhide, Amar - The Venturesome Economy: How Innovation Sustains Prosperity in a More Connected World, 2008
132. Buderi, Robert - Engines of Tomorrow: How the World's Best Companies are Using Their Research Labs to Win the Future, 2000
133. Gertner, Jon - The Idea Factory: Bell Labs and the Great Age of American Innovation, 2012
134. Chesbrough, Henry - Open Innovation: The New Imperative for Creating and Profiting from Technology, 2003
135. Christensen, Clayton - The Innovator's Dilemma: When New Technologies Cause Great Firms to Fail, 2005
136. Utterback, James - Mastering the Dynamics of Innovation, 1994

137. Tassey, Gregory - The Disaggregated Technology Production Function: A New Model of Corporate and University Research, 2005
138. Smith, Douglas and Alexander, Robert - Fumbling the Future: How Xerox Invented, Then Ignored the First Personal Computer, 1999
139. President's Council of Advisors on Science and Technology: Report to the President - Capturing a Domestic Competitive Advantage in Advanced Manufacturing, 2012
140. Gompers, Paul and Lerner, Josh - The Money of Invention: How Venture Capital Creates New Wealth, 2001
141. Thiel, Peter and Masters, Blake - Zero to One: Notes on Startups, or How to Build the Future, 2014
142. Is Silicon Valley Investing in the Wrong Stuff?, Wall Street Journal, 2014
143. https://SBIRsource.com
144. Cole, Jonathan - The Great American University: Its Rise to Prominence, Its Indispensable National Role, Why It must be Protected, 2009
145. Bok, Derek - Universities in the Marketplace: The Commercialization of Higher Education, 2003
146. Kenney, Martin - Biotechnology: The University-Industrial Complex, 1988
147. Etzkowitz, Henry - The Triple Helix: University-Industry-Government Innovation in Action, 2008
148. Stankiewicz, Rikard - Academics and Entrepreneurs: Developing University-Industry Relations, 1986
149. The Economist: Special Report - Companies and the State, February 22, 2014
150. Lerner, Josh - Boulevard of Broken Dreams: Why Public Efforts to Boost Entrepreneurship and Venture Capital have Failed – and What to Do About It, 2012

Chapter 6

151. Wasserman, Noam - The Founder's Dilemmas: Anticipating and Avoiding the Pitfalls That Can Sink a Startup, 2012
152. Moore, Geoffrey - Crossing the Chasm: Marketing and Selling Technology Products to Mainstream Customers, 1991
153. Taleb, Nassim - The Black Swan: The Impact of the Highly Improbable, 2007
154. Friedman, Yali - Building Biotechnology: Business, Regulations, Patents, Law, Politics, Science, 2008
155. The Economist, November 7th, 2015 - Y Combinator: A School for Startups
156. Bloomberg Business Week, October 19-25, 2015 – Investment made by the State of Massachusetts Pays Off
157. The Economist, October 24th, 2015 - The Rise of Startups in Silicon Valley
158. The Economist, October 17th, 2015 - Merger Mania in the Pharmaceutical Industry
159. Lerner, Josh - Boulevard of Broken Dreams: Why Public Efforts to Boost Entrepreneurship and Venture Capital have Failed - and What to Do About It, 2012
160. Eric Ries - The Lean Startup: How Today's Entrepreneurs Use Continuous Innovation to Create Radically Successful Businesses, 2011

Chapter 7

161. Tassey, Gregory - The Technology Imperative, 2009

Chapter 8

162. Tassey, Gregory - Rationales and Mechanisms for Revitalizing U.S. Manufacturing R&D Strategies, 2010
163. KPMG Global Manufacturing Outlook - Preparing for Battle: Manufacturers get Ready for Transformation, 2015

164. Prestowitz, Clyde - The Betrayal of American Prosperity: Free Market Delusions, America's Decline, and How We Must Compete in the Post-Dollar Era, 2010
165. Berger, Suzanne - Making in America: From Innovation to Market, The MIT Task Force on Production in the Innovation Economy, 2013
166. Anderson, Chris - Makers: The New Industrial Revolution, 2012
167. The Economist, November 21st, 2015 - The Industrial Internet of Things: Machine Learning
168. The Economist, November 21st, 2015 - Smart Products, Smart Makers
169. McCormack, Richard - The Plight of U.S. Manufacturing, 2009
170. Nothhaft and Kline - Great Again: Revitalizing America's Entrepreneurial Leadership, 2011
171. Ignite 1.0: Voice of American CEOs on Manufacturing Competitiveness, Council on Competitiveness, 2011
172. Liveris, Andrew - Make It in America: The Case for Re-Inventing the Economy, 2011
173. Pisano, Gary and Shih, Willy - Does U.S. Really Need Manufacturing, Harvard Business Review Special Report, 2012
174. Teixeira et al - Reinventing America: Why the World Needs the U.S. to Bounce Back, Harvard Business Review Special Report, 2012
175. Atkinson, Robert - Why We Need a National Manufacturing Technology Strategy, 2011
176. Shih, Willy and Pisano, Gary - Producing Prosperity: Why America Needs a Manufacturing Renaissance, 2012
177. Global Manufacturing Competitiveness Index, Deloitte, 2013
178. Note: The six startups presented in this chapter highlight manufacturing technologies that, in the first two examples, were created in the U.S. but now are produced or will be produced abroad. The next two examples are technologies being created in the U.S. and potentially can be manufactured here provided the needed industrial commons is not allowed to wither. The last two startup examples highlight innovative, advanced manufacturing technologies being developed and deployed in America.
179. An Assessment of the United States Measurement System: Addressing Measurement Barriers to Accelerate Innovation, National Institute of Standards and Technology, 2007
180. Kaushal, Arvind et al - Manufacturing's Wake-Up Call, 2011
181. Locke, Richard and Wellhausen, Rachel (Editors) - Production in the Innovation Economy, 2013
182. Marsh, Peter - The New Industrial Revolution: Consumers, Globalization and the End of Mass Production, 2012
183. Idelchik, Michael - How the U.S. Can Lead the Next Manufacturing Revolution, 2012
184. PCAST Report to the President on Capturing Domestic Competitive Advantage in Advanced Manufacturing, 2012
185. Rothman, David - Can We Build Tomorrow's Breakthroughs? MIT Technology Review, 2011
186. A Growth Agenda: Four Goals for a Manufacturing Resurgence in America, National Association of Manufacturers, 2012

Chapter 9

187. Kaplan, Jerry - Humans Need Not Apply: A Guide to Wealth and Work in the Age of Artificial Intelligence, 2015
188. Ford, Martin - Rise of the Robots: Technology and the Threat of a Jobless Future, 2015
189. National Academy of Engineering of the National Academy - Grand Challenges for Engineering, 2008
190. https://www.whitehouse.gov/administration/eop/ostp/grand-challenges
191. Atkinson, Robert and Andes, Scott - The Atlantic Century II: Benchmarking EU and U.S. Innovation and Competitiveness, Information Technology and Innovation Foundation, 2011

192. McKinsey Global Institute - An Economy That Works: Job Creation and America's Future, 2011

193. Note: The underlying theme for the six examples presented in this chapter is to highlight the potential of technologies being developed by taxpayer-funded startups to seed whole new industries in America.

194. PCAST Report to the President on Capturing Domestic Competitive Advantage in Advanced Manufacturing, 2012

195. Business Week, November 2015 - Maker's Row, Help for Small Business Manufacturing Needs

196. Katz, Bruce and Bradley, Jennifer - The Metropolitan Revolution: How Cities and Metros are Fixing Our Broken Politics and Fragile Economy, 2013

197. The Economist, August 08, 2015 - Intellectual Property: A Question of Utility

198. Boldrin, Michele and Levine, David - The Case Against Patents, 2012

199. Johns, Adrian – Piracy: The Intellectual Property Wars from Gutenberg to Gates, 2010

200. The Economist, October 24th, 2015 - Reinventing the company: Entrepreneurs are redesigning the basic building block of capitalism

201. Susskind, Richard and Susskind, Daniel - The Future of the Professions: How Technology Will Transform the Work of Human Experts, 2016

202. Morgan, Jacob - The Future of Work: Attract New Talent, Build Better Leaders, and Create a Competitive Organization, 2014

203. Thompson, Derek - A World Without Work, The Atlantic, July/August 2015

204. Jackson, Andrew and Dyson, Ben - Modernizing Money: Why our Monetary System is Broken and How it Can be Fixed, 2013

205. Ross, Carne - The Leaderless Revolution: How Ordinary People Will Take Power and Change Politics in the 21st Century, 2012

206. Nader, Ralph - Unstoppable: The Emerging Left-Right Alliance to Dismantle the Corporate State

207. Rifkin, Jeremy - The Zero Marginal Cost Society: The Internet of Things, the Collaborative Commons, and the Eclipse of Capitalism, 2015

Index

A

A123 Systems, 79
Additive manufacturing, 6, 183, 202, 222
Adhesion, 183
Advanced batteries, 12, 57, 181, 202
Advanced Manufacturing Cooperatives, 196
Aleutian Arc, 70
Allergan, 143
Amazon, 47, 109, 150, 206
American Institute of Chemical Engineers, 138
Amniotic membrane, 52
Angel Capital Association, 137, 200
Apollo, 65
App economy, 150
Apple, 11, 47, 59, 113, 141, 168, 177, 201
Application specific integrated circuit, 102
Aquaculture, 157
Arizona State University, 68
Artificial creatures, 26
Artificial Intelligence, 3, 29, 57, 186, 190, 209, 229
Artificial life, 5, 26, 31, 34, 209
Artificial reality, 29, 190
Artificial satellite, 65
Artificially provoked neo-life, 34
Asteroid Grand Challenge, 187
Asteroid Redirect Mission, 66
Atomic force microscope, 105
Atomistic disorder, 71
Atrial fibrillation, 134
Autistic spectrum disorders, 62
Automated manufacturing, 14
Automated voice recognition, 72
Autonomous wearable devices, 90

B

Bacon, 11, 18, 19, 224
Balance of payments, 167
Bangalore, 27, 119, 149
Battery manufacturing, 9, 177
Bayh-Dole, 52, 115
Baylor College of Medicine, 206
B-Corp movement, 211
Beam-steering, 93
Bell Labs, 25, 47, 59, 102,103, 225, 227
Benefit Corporation, 212
Bio-based economy, 89
Bio-chip, 6
Bio-composites, 99
Bio-computing, 53, 209
Biodegradable Biopolymer, 33, 219
Bioethics, 22
Bio-fabrication, 12
Bio-informatics, 209
Biological humanity, 17
Biologically-inspired adhesives, 193
Bioluminescent organisms, 71
Bio-manufacturing, 138, 181, 182, 209
Biomarker, 62, 160

Biomaterials, 70, 132, 150, 160, 179, 221
Biometric analysis and identification, 22
Biomimetics, 162, 190, 209, 222
Bionic devices, 156
Bioplastics, 33
Bio-refineries, 5
Biotechnology Age, 14
Biotechnology Industry Organization, 137, 138
Bitcoin, 210
Block chain technology, 210
Blockbuster drugs, 143
Boil-off, 86
Bone Densitometer, 49
Bose-Einstein condensate, 69
Botox, 143
Brain biochemistry, 62
Brokering, 43, 59, 103
Bucky Balls, 78
Build-test-design cycle, 185
Built-in obsolescence, 168
Burden of knowledge, 1, 7
Business intelligence, 192
Buzz Aldrin, 65

C

California Institute of Technology, 68
Carbon fiber, 29
Carbon footprint, 161, 204
Carbon-free emissions, 7
Cardiac patch, 4
Care-improvement algorithms, 211
Carnegie Mellon University, 193
Catheter-related infections, 30
Cellular telephony, 42
Center for Advanced Materials Characterization, 90
Cervical pap-smear, 7
Chemical etching, 193
Chemical vapor deposition, 184
CHiP Lube system, 108
Circular economy, 215
Cisco, 49, 59, 103, 109
Climate change, 2, 11, 63, 66, 80, 85, 161, 162, 188, 190, 208
Clock tree, 101
Cloning, 6, 22, 26
Cloud Computing, 77, 210
Cloud-based, 28, 114
Cognition drugs, 6
Cognitive automobiles, 29
Cognitive radios, 28
Coherent neutrino-nucleus scattering, 64
Coherent regenerator matrix, 86
Cold war, 24
Collaborative commons, 212, 214, 230
Compaq, 113
Comparative advantage, 81, 92
Comparative effectiveness research, 211
Competitive advantage, 8, 14, 39, 40, 41, 109, 165, 167, 181, 195, 196, 225, 228, 229, 230
Competitiveness index, 173, 229
Complementary metal-oxide semiconductor, 96
Component-supplier model, 90

Compressed-air energy storage, 161
Computer vision, 24, 121
Computer-aided design, 78, 122
Consortia, 136, 139, 172, 174, 208
Consumer Electronics Show, 138
Consumption smoothing, 161
Corporate innovation models, 138, 192
Corporate venture capital, 103, 104, 111, 138, 140, 200
Cradle-to-cradle, 33, 182
CRISPR/Cas9 system, 210
Cross-licensing, 97, 126
Crowd-funding, 14, 199, 200, 210, 211
Cryogenic pyro-valves, 51
Cryptography, 210
Crystalline silicon wafers, 181
Customer discovery process, 131
Cyber-infrastructure, 39, 209
Cyber-physical systems, 28, 184
Cyber-security, 74

D

Decellularization, 4
Defense Advanced Research Projects Agency, 25, 58, 72, 76
Defense Production Act, 87
Deoxyribonucleic acid, 26, 115
Department of Energy, 48, 63, 76, 79, 81
Department of Innovation, 208
Designed-for-manufacturability, 50
Designed-for-serviceability, 50
Detector technologies, 64
Development bank, 180, 212
Developmental disabilities, 62
Diffusion, 12, 31, 39, 48, 77, 97, 116, 180
Digital Revolution, 75
Digital subscriber lines, 93
Digitization of life, 26
Digitization of work, 32
Directional antennas, 93
Discovery science, 5, 14, 55
Discovery-to-innovation-to-market, 190
DNA origami, 6
DNA strands, 26
Do-it-yourself biology, 6, 209
Doppler radar, 78
Dreamit Ventures, 142
Dry adhesive, 193

E

Early adopters, 14, 91, 98, 99, 100, 125, 131
Earth Observing System, 65
Eastman Business Park, 137, 152, 153, 154, 155, 194
Economic determinism, 32
Economic force multiplier, 169
Elastomeric, 193
Electric vehicle, 79, 80, 182, 187, 191, 202
Electric-powered wheelchair, 121
Electrochemical compressor, 76, 81, 220
Electromagnetic interference mitigation, 88
Electromyography signals, 213

Electron Spin Resonance, 159
Electronic Design Automation, 57, 100, 101, 221
Electronic medical records, 211
Electronics recycling, 50
Electrostatic surface haptics, 117
Embedded object recognition, 121
Emoticons, 14
Emotion norms database, 114
Emotion-reading machines, 114, 221
Endogenous growth theory, 38
End-user innovation, 44
Energy arbitrage, 161
Energy storage, 9, 27, 29, 139, 150, 161, 162, 175, 182, 184, 191, 219, 222
Energy-intensive plastics, 33
Engineered microbial strains, 214
Engineering Research Centers, 69, 70, 194, 217
Environmental remediation, 32
Enzyme designers, 14
Epigenetics, 6, 209
Estrogenic activity, 23, 219
Evidence-based measures, 151
eWaste, 50
Exascale supercomputer, 28
Export-Import Bank, 198, 212
Exteroceptive sensing, 121

F

Factor of production, 12, 39
Fail-safe mechanism, 50
Fair return model, 117
Family offices, 213
Federal Drug Administration, 126
Federal Reserve, 202, 210
Feedstock, 22, 33, 99, 182, 209, 214
Femtosecond laser technology, 84
Field programmable gate array, 102
Financial Industry Regulatory Authority, 156
Fire-retardant foams, 99
First-in-class biologic, 77
First-to-file, 97
First-to-invent, 97
Flexible displays, 153
Flexible electronics, 9,175, 181, 202, 209
Flight critical software, 120
Fluorescent, 71
Fluorinated photoresist system, 153
Force-sensitive touch, 174
Formaldehyde-free adhesives, 99
Frequency-modulated radio, 115
Functional thin-films, 150

G

Galaxy Evolution Explorer, 67
Garage-to-market, 15, 124, 132, 150
Gas ionization detectors, 65
Gemini, 65
Gene therapy, 96, 160
Genentech, 103, 109, 111, 143
General-purpose technology, 125

Genetic anlage, 62
Genetically-modified, 14
Genetics, 5, 6, 26, 57, 69, 209
Genome editing, 115, 210
Genzyme, 103
Geographic Information Systems, 85
Georgia Tech Research Corporation, 141
Geothermal tracers, 13
Glioblastoma multiforme, 77
Global Positioning System, 57, 71, 72, 73, 78, 115
Global warming, 5, 33, 34
Globalization, 1, 11, 35, 97, 119, 172, 180, 205, 224, 227
Google Glass, 146
Grand Challenges for Development, 187
Graphene synthesis, 184
Graphical User Interface, 42, 50
Gravitational wave observatories, 69
Great-circle routes, 71
Greenhouse-gas emissions, 33
Grid Energy Router, 191
Gross Domestic Product, 40, 43, 168, 169, 170
Ground-level knowledge, 45, 100
Guangdong, 119

H

H5N1 bird flu, 74
Haptic touchscreens, 117
Hardware service providers, 14
Harvard Medical School, 142
Headset controller device, 213
Health information specialists, 61
Heidegger, 18, 32, 224, 225
Hemodialysis, 30
Hernia repair, 4, 179
Hierarchical classifier, 24
Hollowed-out, 13, 126, 175, 190
Hub-and-spokes, 91
Hubble Space Telescope, 66
Hubble Ultra Deep Field, 67
Human assistive technology, 213
Human cloning, 6
Human sperm-making biological machine, 28
Human-computer interface research, 59
Human-in-the-loop cooperative control, 121
Hybrid creatures, 26
Hyper-connected, 32, 196

I

IBM, 35, 47, 103, 138, 140, 206
Immunotherapy, 157
Impact investing, 213
Increasing-returns industries, 50
Industrial biotechnology, 5
Industrial commons, 11, 24, 86, 124, 158, 165, 170, 173, 197, 229
Industrial environmental monitoring, 80
Industrial microbiology, 137, 139, 175, 182, 190, 197, 200, 209
Industrial Research Institute, 137, 138
Industrial Revolution, 55, 75, 167, 180, 186, 205, 229
Industry-recognized skills credentials, 183

Industry-university partnerships, 189
Information Age, 14, 191
Information security ratings, 151
Infrared chemical nanospectroscopy, 105
Infra-technologies, 85, 124, 158, 160, 208
Ingestible chips, 5
Initial public offering, 79, 132, 148, 154, 212
Innovation capacity, 13, 187
Innovator-entrepreneur, 47, 188
Institutes for Manufacturing Innovation, 197
Instrumentation and control, 126
Integrated circuit fabrication, 96
Integrated photonics, 181, 190, 209
Integrated voltage regulators, 175
Integrated-circuit design, 101
Intel Capital, 174
Intellectual property, 6, 8, 12, 14, 38, 43, 77, 91, 95, 96, 102, 103, 117, 119, 122, 126,
            127, 129, 130, 131, 133, 135, 138, 144, 148, 151, 152, 192, 193,
            196, 201, 204, 230
Intellectual property exchange platform, 130, 133
Intellectual property law, 38
Intelligent materials, 6, 209
Intercloud, 3
Interface chemistry, 183
International Finance Facility for Immunization, 210
International Labor Organization, 205
International Space Station, 49, 65
Internet of things, 3, 74, 167, 168, 182, 190, 209, 213, 229, 230
Interoperability protocols, 85, 158
iPhone, 57
Ischemic, 4

J

James Webb Space Telescope, 66, 68
Journey to Mars, 66
Jumpstart Our Business Startups (JOBS) Act, 213
Juno spacecraft, 66

K

Kauffman Foundation, 52, 94, 128, 135, 137, 226, 227
Keyhole Limpet Hemocyanin, 157
Keynesian economics, 37
Kickstarter, 142, 206, 213
Knowledge integration toolkit, 207
Knowledge-based innovation, 22, 76, 94, 208

L

LabCentral, 141, 142
Lab-scale prototype, 150, 194
Lab-to-manufacture, 15, 124, 132, 150, 152, 153, 194
Laissez-faire, 12, 57, 226
Large neutrino magnetic moments, 64
Large-area optical films, 204
Laser, 7, 13, 24, 48, 78, 84, 115, 125, 181, 189
Lean manufacturing, 82
Lean startup, 82, 145, 227, 228

Life Sciences Act, 141
Light emitting diodes, 10, 29
Lignocellulosic, 99
LinkedIn, 133, 134, 141
Liquid Crystal Displays, 153, 171, 181
Lithium ion batteries, 79
Location-aware sensors, 90
Location-based services, 57
Low-carbon manufacturing, 166
Low-viscosity resins, 183
Lucent's New Ventures Group, 104
Luminal infections, 30

M

Machine learning, 24, 32, 129, 168, 186, 219, 229
Machine tools, 37, 164
Machine-to-machine, 75, 182
Magnetic Resonance Imaging, 78, 115
Maker movement, 6, 13, 167
Maker's Row, 197, 198, 23
Makerspaces, 196
Manufacturing Extension Partnership, 136, 137
Manufacturing Institute, 155
Manufacturing supply-chain, 136, 165
Mars Pathfinder, 65
Massive Open Online Courses, 162, 207
Matching logic verification, 120
Matter compilers, 6
Medical libraries, 61
Medical-grade materials, 49
Megathura crenulata, 157
Mercury, 65
Meta-materials, 73, 74
Metastability, 101
Metrology, 126
Metropolitan export plans, 198
Microbusiness innovation science and technology survey, 70
Microelectromechanical systems, 174
Microfluidic valve system, 159
Micro-lending, 5
Micro-projectors, 175
Microsoft, 47, 109, 146, 149, 150, 201
Micro-spectrometer, 159, 222
Military R&D, 57
Military-industrial complex, 24
Minimally invasive surgery, 144
Minimum quantity lubrication, 108, 221
Miracle drugs, 19
MIT Lincoln Labs, 150
MIT Venture Mentoring Service, 137
Mobile device power management, 126
Molecular and cellular biology, 26
Molecular sequencing, 211
Mosaic, 136
Motorola, 47, 201
Multiple clock domains, 101
Multiple input multiple output radio technology, 93
Multiplier effect, 40

N

Nano-bio-info convergence, 5
Nanobots, 27
Nanoelectronics, 28
Nanotechnology, 12, 13, 29, 34, 57, 58, 59, 77, 87, 97, 105, 126, 174, 181, 182, 190, 209
NASA Commercial Readiness Program, 49
National Academy of Engineering, 187, 229
National Advisory Committee on Aeronautics, 65
National Aeronautics and Space Administration, 65, 76
National Association of Manufacturers, 136, 138, 183, 229
National Export Initiative, 199
National Innovation Council, 208
National Institutes of Health, 56, 58, 60, 137, 146
National Laboratories, 63, 116
National Network for Manufacturing Innovation, 196
National Science Foundation, 4, 8, 58, 68, 76, 77, 83, 89, 216
National Venture Capital Association, 137, 200
Navy nuclear propulsion program, 63
Near-zero marginal cost, 213
Negative externalities, 53
Neil Armstrong, 65
Network effects, 140
Neurodevelopmental Disabilities, 62
Neuroscience, 3, 26, 160
Neutrino physics, 64
New Horizons spacecraft, 66
New Industries Endowment, 208
New Industries Foundation, 208
Newton, 18, 195
Next-generation internet search, 129, 221
Next-generation wearable controller, 213
Nobel Prizes, 69
Non-germicidal filtration systems, 10
Non-lithium materials, 9
Non-volatile memory, 96
Nortel Networks, 201
Not-invented-here syndrome, 103
Nuclear and radiological incidents, 63
Nuclear deterrent without testing, 63
Nuclear weapons production, 63

O

Objectification of children, 34
Ocean acidification, 34
Oculus VR, 142
Office of Naval Research, 88
Oncology diagnostics, 77
One-transistor memory device, 96, 221
Open hardware movement, 6,196
Open source movement, 98
Ophthalmic wound healing, 52
Optical fibers, 24
Optoelectronics, 12, 13, 37, 53, 57, 127, 131, 175, 176, 181, 187, 190, 197, 200, 202
Oracle, 103
Oregon Nanoscience and Microtechnologies Institute, 90
Organic electronic devices, 153
Organic Light Emitting Diode, 153
Organism engineering foundry, 214
Orion spacecraft, 66

Orphan Drug Act, 56
OTC Markets Group, 156
Ozone hole, 78

P

Pacific Northwest National Laboratory, 64, 90
Particle physics, 65
Patent reform law, 97
Peer-to-peer lending, 210, 212
Perfusion, 4
Peri-auricular muscles, 213
Peritoneal dialysis, 30, 219
Personalized medicine, 5, 49, 75, 162, 190, 209, 211
Petrochemical-based polymers, 33
Petroleum-based plastics, 33, 99
Pfizer, 47, 143
Pharmaco-genetics, 5
Photolithographic infrastructure, 153
Pilot-scale manufacturing, 107
Plant identification system, 24
Plant science, 80
Polyhedral Oliogomeric Silsesquioxane, 87
Polyhydroxyalkanoate, 33
Polymer electrolyte, 9
Portable photosynthesis system, 76, 80, 220
Positive externalities, 53, 119
Post-industrial manufacturing ecosystem, 166
Post-petroleum age, 7
Post-work society, 208
Power management integrated circuits, 175
Powered knee-ankle prosthetic, 156, 222
Precision medicine, 5
Precision weapons, 72
Precision agriculture, 49, 209
President's Council of Advisors on Science and Technology, 109, 194, 228
President's national innovation advisor, 208
Printed circuit boards, 164, 171,175
Print-on-demand, 26
Programmable materials, 28
Programming language of life, 45
Proof-of-concept, 119, 120, 121, 124, 152
Proteolysis-based tools, 214, 223
Proteomics, 26, 160, 162, 209
Public-good content, 119
Publicly-funded laboratories, 37, 118
Public-private partnerships, 13, 25, 124, 143, 174, 211
Pulse tube cryocooler, 86

Q

Qualcomm, 80, 84, 113
Quantum assisted sensing, 73
Quantum compass, 73
Quantum computing, 28, 162, 175, 182, 190, 209
Quantum dots, 13, 72, 219

R

R&D Magazine, 77

Radio frequency identification, 22, 78
Random access, 19, 96
Rapid business model prototyping, 42
Rapid threat assessment, 74
Rare earth alloys, 86
Real-time on-orbit analysis, 49
Recombinant protein, 89
Recycling, 5, 50, 209, 220
Regenerative biologic therapies, 52, 220
Regenerative medicine, 6, 26, 162, 190, 209
Regional innovation clusters, 139
Renewable energy, 3, 29, 64, 119, 131, 161, 162, 170, 177, 191, 213
Repositories, 133, 211
Respired hydrogen, 28
Reusable substrate, 184
Revascularized, 4
Revenue-based financing, 212
Reverse engineering, 6
Right-of-first-refusal, 104
Ring of fire, 70
Riparian Pharmaceuticals, 142
Risk management, 151
Robobowl, 194
Robotic exoskeletons, 156
Rotator cuff tears, 179
Runtime verification, 120

S

Satellites, 24, 57, 65
Scalable semantics-based, 120, 221
Scaling bank, 180
Schumpeter, 32, 38, 41, 92, 225, 227
Science and engineering indicators, 69, 227
Scientist robots, 31
Second machine age, 75, 227
Second valley of death, 111, 197
Secretary of Innovation, 208
Self-programming thinking machines, 29
Self-replication, 34
SEMATECH, 138
Semi-autonomous cooperative navigation, 121
Semiconducting polymers, 153
Shape memory polymers, 179
Shared-use facilities, 197
Siemens, 139, 140
Silicon Valley, 14, 41, 48, 113, 119, 140, 142, 149, 202, 205, 228
Silicon-IP, 102
Singapore, 119
Skin-to-air interface, 90
Small- and medium-sized enterprises, 136, 197, 199
Small Business Innovation Development Act, 52, 83
Small Business Reauthorization Act, 84
Small interfering ribonucleic acid technology, 210
Smart electricity grids, 29
Smart materials, 182
Smart-window technology, 204
Social entrepreneurship, 5
Social-emotional intelligence, 114
Socialization of losses, 59
Soft tissue reinforcement device, 179
Software-as-a-service, 114

Solid-state lighting, 29, 162, 190, 209
Solid-state physics, 18
Solid-state transformer, 191, 222
Space Launch System, 51, 66
Space shuttle, 65
Space weather, 66
Spallation Neutron Source, 64
Specific field-of-use, 119
Specific impulse, 86
Spectrum usage, 97
Spiral welding, 178, 222
Stanford University, 102, 141, 194
StartX, 137, 141, 194
Static random access memory, 96
Stealth technology, 72
Stereolithography, 183
Sterile neutrinos, 64
Steve Jobs, 205
Stone-age biology, 17
Storage capacity, 28, 161
Submarine patents, 203
SunShot, 82, 187
Superconducting quantum interference device, 88
Supercritical carbon dioxide, 108
Supply-chain, 11, 13, 36, 85, 99, 107, 124, 126, 130, 133, 136, 138, 151, 158, 165, 166, 167, 175, 177, 180, 190, 192, 196, 197, 200, 208
Supreme Court, 207
Surgical instruments, 144, 209
Sustainable development challenges, 152
Synthetic biology, 5, 6, 13, 34, 57, 70, 76, 89, 114, 160, 162, 175, 182, 190, 209, 220
Synthetic brain subsystems, 27
Synthetic life, 5, 6
Synthetic organ replacement, 87
Synthetic spider silk fibers, 89, 220
System-on-a-chip, 96

T

Tapered spiral welded tower, 178
Technical civilization, 17
Techno-eugenic rat race, 34
Technology development curve, 124
Technology Enhancement for Commercial Partnerships, 88
Technology in-licensing, 138
Technology landscape mapping, 192
Technology licensing offices, 135, 192, 193
Technology life-cycle transitions, 187
Technology roadmaps, 122, 138
Technology spillovers, 53
Technology standards, 97
Technology to Business Centers, 139
Technology-specific networks, 14, 91, 100, 124, 131, 192
Techstars, 142
Tech-transfer offices, 118, 122
Tel Aviv, 119
Terahertz, 73, 74
Thermoelectric generator, 90
Thickets of patents, 204
Three-dimensional bio-printer, 6
Three-dimensional imaging, 121
Three-dimensional printing, 12, 27
Timing closure, 101

Tissue engineering, 49, 78, 160, 190, 209
Tissue regeneration, 179
Toll manufacturing, 195
Total factor productivity, 36
Touch-based human interface technology, 174
Traffic congestion charging, 7
Trans-femoral prostheses, 156
Transistor, 24, 42, 71, 72, 96, 153, 221
Transmission congestion, 161
Transmission electron microscopy, 48, 184
Trans-Pacific Partnership, 203
Transparent conductive coatings, 93
Triple helix, 13, 116, 228
Tunneling, 71

U

U.S. biotech industry, 56
U.S. Patent and Trademark Office, 8, 98
U.S. Small Business Innovation Corporation, 113
Ultra-accurate motion sensing, 175
Ultraviolet coverage, 67
Umbilical cord-based tissue products, 52
Under-utilized manufacturing facilities, 109, 139, 155
Unicorns, 140, 206
University of Cincinnati, 52
University of Columbia, 52
University of Manchester, 52
University of Massachusetts, 142
University of Miami, 52
University of Nebraska, 80
University of Oregon, 90
University of Pennsylvania, 62
University of Texas, 52
UNIX operating system, 42
Upwork, 206

V

Validated learning, 82
Valuation, 50, 85, 104, 111, 120, 128, 129, 145, 154, 158, 179, 211, 213
Value-added manufacturing, 166
Variable-gravity platform, 49
Vascular, 4
Vector machine, 24
Vendor supply-chain, 151
Venturesome customers, 14, 91
Vertically-integrated, 12, 46, 102, 131, 174
Vestigial, 213, 223
Virtual-reality, 142
Visual object identification, 24
Vizio, 206
Voice-activated personal assistant, 57
Volatile Organic Compounds, 87

W

Wanxiang Group, 80
Water recycling, 5
Weapons of mass destruction, 63

Wearable electronics, 9
Wearable energy harvester, 76, 90, 220
Wearable robots, 156
Web browser, 78, 167
WebMD, 207
Wet chemical synthesis, 13
WeWork, 206
WhatsApp, 112, 206
Wheel of innovation, 91
Wireless mesh network, 93
World Trade Organization, 180

X

Xerox, 46, 47, 102, 107, 228

Y

Y Combinator, 140, 141, 217, 228
Yahoo!, 109, 146

Z

Zero marginal cost, 75, 213, 230
Zero-emissions storage solution, 161
Zinc battery chemistry, 9

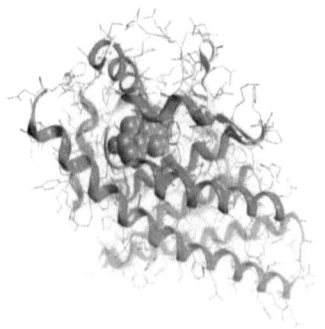